Ocean Currents

Teacher's Guide

Grades 5–8

Skills

Observing • Inferring • Making Models • Visualizing • Explaining • Comparing
Recording • Working Cooperatively • Communicating • Logical Thinking
Critical Thinking • Drawing Conclusions

Concepts

Oceanography • Geography • Ocean Currents • Pollution
Waste Disposal • Density • Wind • Salinity • Temperature
Upwelling and Downwelling Zones •
Impact of Currents on Human History •
Importance of the Ocean to the Planet and Daily Life

Themes

Models and Simulations • Systems and Interactions • Patterns of Change

Mathematics Strands

Number • Measurement • Pattern • Logic and Language

Nature of Science and Mathematics

Creativity and Constraints • Interdisciplinary • Real-Life Applications

by

Catherine Halversen
Kevin Beals
Craig Strang

ceanCurrentsOceanCurrentsOcean

Great Explorations in Math and Science
Lawrence Hall of Science
University of California at Berkeley

Lawrence Hall of Science
University of California at Berkeley

Director
Ian Carmichael

Cover Design and Illustrations
Lisa Haderlie Baker

Publication of *Ocean Currents* was made possible by a grant from the Employees Community Fund of Boeing California and the Boeing Corporation (originally the McDonnell-Douglas Foundation and Employees Community Fund). The GEMS Program and the Lawrence Hall of Science greatly appreciate this support.

COMMENTS WELCOME

Great Explorations in Math and Science (GEMS) is an ongoing curriculum development project. GEMS guides are revised periodically, to incorporate teacher comments and new approaches. We welcome your criticisms, suggestions, helpful hints, and any anecdotes about your experience presenting GEMS activities. Your suggestions will be reviewed each time a GEMS guide is revised. Please send your comments to:

GEMS Revisions
Lawrence Hall of Science
University of California
Berkeley, CA 94720-5200

Phone: (510) 642-7771

Fax: (510) 643-0309

gems@uclink4.berkeley.edu

www.lhsgems.org

Initial support for the origination and publication of the GEMS series was provided by the A.W. Mellon Foundation and the Carnegie Corporation of New York. Under a grant from the National Science Foundation, GEMS Leader's Workshops have been held across the country. GEMS has also received support from: the McDonnell-Douglas Foundation and the McDonnell-Douglas Employee's Community Fund; Employees Community Fund of Boeing California and the Boeing Corporation; the Hewlett Packard Company; the people at Chevron USA; the William K. Holt Foundation; Join Hands, the Health and Safety Educational Alliance; the Microscopy Society of America (MSA); the Shell Oil Company Foundation; and the Crail-Johnson Foundation. GEMS also gratefully acknowledges the contribution of word processing equipment from Apple Computer, Inc. This support does not imply responsibility for statements or views expressed in publications of the GEMS program.

For further information on GEMS leadership opportunities, or to receive a catalog and the *GEMS Network News*, please contact GEMS at the address and phone number opposite. We also welcome letters to the *GEMS Network News.*

Printed on recycled paper with soy-based inks.

International Standard Book Number: 0-924886-44-7

Great Explorations in Math and Science (GEMS) Program

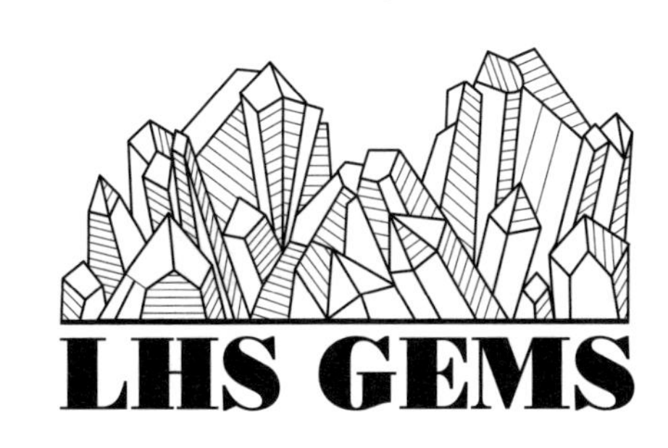

The Lawrence Hall of Science (LHS) is a public science center on the University of California at Berkeley campus. LHS offers a full program of activities for the public, including workshops and classes, exhibits, films, lectures, and special events. LHS is also a center for teacher education and curriculum research and development.

Over the years, LHS staff have developed a multitude of activities, assembly programs, classes, and interactive exhibits. These programs have proven to be successful at the Hall and should be useful to schools, other science centers, museums, and community groups. A number of these guided-discovery activities have been published under the Great Explorations in Math and Science (GEMS) title, after an extensive refinement and adaptation process that includes classroom testing of trial versions, modifications to ensure the use of easy-to-obtain materials, with carefully written and edited step-by-step instructions and background information to allow presentation by teachers without special background in mathematics or science.

Staff

Director: Jacqueline Barber

Associate Director: Kimi Hosoume

Associate Director/Principal Editor: Lincoln Bergman

Mathematics Curriculum Specialist: Jaine Kopp

GEMS Network Director: Carolyn Willard

GEMS Workshop Coordinator: Laura Tucker

Staff Development Specialists: Lynn Barakos, Katharine Barrett, Kevin Beals, Ellen Blinderman, Gigi Dornfest, John Erickson, Stan Fukunaga, Debra Harper, Linda Lipner, Karen Ostlund

Distribution Coordinator: Karen Milligan

Workshop Administrator: Terry Cort

Financial Assistant: Vivian Tong

Distribution Representative: Felicia Roston

Shipping Assistant: Maureen Johnson

GEMS Marketing and Promotion Director: Matthew Osborn

Senior Writer: Nicole Parizeau

Senior Editor: Carl Babcock

Editor: Florence Stone

Director of Publications: Kay Fairwell

Art Director: Lisa Haderlie Baker

Senior Artists: Carol Bevilacqua, Rose Craig, Lisa Klofkorn

Staff Assistants: Haleah Hoshino, Mikalyn Roberts, Thania Sanchez, Stacey Touson

Contributing Authors

Jacqueline Barber
Katharine Barrett
Kevin Beals
Lincoln Bergman
Susan Brady
Beverly Braxton
Kevin Cuff
Linda De Lucchi
Gigi Dornfest
Jean Echols
John Erickson
Philip Gonsalves
Jan M. Goodman
Alan Gould
Catherine Halversen
Debra Harper
Kimi Hosoume
Susan Jagoda
Jaine Kopp
Linda Lipner
Larry Malone
Cary I. Sneider
Craig Strang
Herbert Thier
Jennifer Meux White
Carolyn Willard

Reviewers

We would like to thank the following educators who reviewed, tested, or coordinated the reviewing of this group of GEMS guides (*Ocean Currents, Only One Ocean,* and *Environmental Detectives*). Their critical comments and recommendations, based on classroom presentation of these activities nationwide, contributed significantly to this publication. Their participation in this review process does not necessarily imply endorsement of the GEMS program or responsibility for statements or views expressed. Their role is an invaluable one; feedback is carefully recorded and integrated as appropriate into the publications.

THANK YOU!

ARIZONA

Sonoran Sky Elementary School, Scottsdale
Brent Engilman
Marge Maceno *
Amy Smith
Kathy Wieeke
Tammy Wopnford

ARKANSAS

Bob Courtway Middle School, Conway
Robin Cole
Rick Hawkins
Charlcie Strange *
Paula Wilson

Birch Kirksey Middle School, Rogers
Beth Ann Carnes
Jenny Jones *
Curtis S. Smith
Sharron Wolf

CALIFORNIA

Albany Middle School, Albany
Karen Adams
Jenny Anderson
Cyndy Plambeck
Kay Sorg

Martin Luther King Middle School, Berkeley
Indigo Babtiste
Akemi Hamai
Yvette McCullough
Beth Sonnenberg

Harding School, El Cerrito
Renie Gannett
Carol Leitch
Jim Wright

Portola Middle School, El Cerrito
Debbie Marasaki
Carol Mitchell
Susan Peterson
Mike Wilson

Warwick Elementary School, Fremont
Dale Harden
Katy Johnson
Richard Nancee
Robert Nishiyam
Bonnie Quigley
Ann Trammel

Ohlone Elementary School, Hercules
Stacey Cragholm
Gloria Crim
Jay Glesener
Sandra Simmons

Hall Middle School, Larkspur
Trish Mihalek
Art Nelson
Ted Stoeckley
Barry Sullivan

McAuliffe Middle School, Los Alamitos
Michelle Armstrong
Kathy Burtner

Oak Middle School, Los Alamitos
Joyce Buehler
Rob Main *
Kendall Vaught

Hidden Valley School, Martinez
Diane Coventry
Nigel Dabby
Jennifer Sullivan

Calvin Simmons Jr. High School, Oakland
Stan Lake
Wendy Lewis
Fernando Mendez
Thelma Rodriguez

Vintage Parkway School, Oakley
Jennifer Asmussen
Alisa Haley
Casey Maupin
Lian McCain
Steve Williams

Collins Elementary School, Pinole
Ralph Baum
Craig Payne
Anne Taylor
Genevieve Webb

Adams Middle School, Richmond
Richard Avalos
Susan Berry
John Eby
John Iwawaki
Steve Stewart

*** Trial Test Coordinators**

Bell Jr. High School, San Diego
Nick Kardouche
Elouise King *
Denise Vizcarra
Mala Wingerd

Dingeman Elementary, San Diego
Monka Ely *
Godwin Higa
Kim Holzman
Linda Koravos

Cook Middle School, Santa Rosa
Steve Williams

Rincon Valley Middle School, Santa Rosa
Sue Lunsford
Penny Sirota *
Laurel VarnBuhler

FLORIDA

Howard Middle School, Orlando
Elizabeth Black
Carletta Davis
Susan Leeds *
Jennifer Miller

MISSOURI

St. Bernadette School, Kansas City
Brett Coffman
Dorothy McClung
Aggie Rieger
Margie St.Germain

Poplar Bluff 5th-6th Grade Center, Poplar Bluff
Cindy Gaebler
Leslie Kidwell
Barbara King *
Melodie Summers

NEVADA

Churchill County Jr. High School, Fallon
Kerri Angel
Deana Madrasco
Amy Piazzola
Sue Smith-Ansotegui *

NEW HAMPSHIRE

Crescent Lake Elementary, Wolfeboro
Kate Borelli
Amy Kathan
Elaine M. Meyers *
Patti Morissey

NEW JERSEY

Orchard Hill School, Skillman
Jay Glassman *
Al Hadinger
Georgiana Kichura
Tony Tedesco

NEW YORK

Maple Hill Middle School, Castleton
Beth Chittendo
Jeanne Monteau *

Lewisboro Elementary School, South Salem
Debra Jeffers

St. Brigid's Regional Catholic School, Wateruliet
Patricia Moyles

OHIO

Baker Middle School, Marion
Dave Dotson
Denise Z. Iams *
Betty Oyster
Carol White

OREGON

Sitton School, Portland
David Lifton
Deborah Nass

TEXAS

Colleyville ISD, Grapevine
Kathy Keeney

Grapevine ISD/Administration Building, Grapevine
Shelly Castleberry
Terry Dixon
Malanie Gable*
Randy Stuempfig

Ector County ISD, Odessa
Becky Stanford

Ireland Elementary School, Odessa
Susan Hardy

Miliam Elementary School, Odessa
Eli Tavarez

Travis Elementary School, Odessa
Patty Calk
Stacey Hawkins

WASHINGTON

The Gardner School, Vancouver
Matt Karlsen
Tom Schlotfeldt
Rob VanNood

Acknowledgments

The authors want to thank **Tom Murphree** from the Naval Postgraduate School in Monterey, California, for so generously sharing his time and expertise in his thorough scientific review of this guide. His knowledge, insight, and edits over the years have been extremely helpful. Tom's expertise and his ability to work with educators to instill excitement and impart an understanding of the physical processes setting the ocean in motion has been invaluable in the development of *Ocean Currents*.

Thanks to **Dr. E.C. Haderlie**, Distinguished Professor of Oceanography at the Naval Postgraduate School for his willingness to read through *Ocean Currents* with an expert's eye toward assuring its scientific accuracy. Additionally, the authors feel fortunate and thankful to Dr. Haderlie for writing the foreword to this guide.

We thank the authors of the GEMS guide *Discovering Density* (**Jacqueline Barber, Marion E. Buegler, Laura Lowell,** and **Carolyn Willard**). We based Activity 4, Layering Liquids, in this guide on an activity in *Discovering Density.* Thanks as well to **Sarah Christie**, for early encouragement and advice on several activities.

Finally, we thank the students of Ohlone Elementary School in Hercules, California, along with their teacher, **Sandra Simmons,** for testing this unit and providing us with the wonderful photographs of students in action throughout this guide. Photos of students on pages 70 (top) and 81 were taken by **Richard Hoyt.** Photos of students on pages 28, 70 (bottom), 84, 98, and 151 were taken by **Carl Babcock.** The photographs of ships on pages 94 and 104 were taken by **Craig Strang.**

Special filial thanks to retired U.S. Navy Commander **Donald Beals** who inspired a fascination with all things aqueous in one of the authors, in part through a constant willingness to share his vast knowledge of Essex seafaring history.

Contents

Foreword

It is a pleasure to write a brief foreword to this latest LHS GEMS Teacher's Guide. As with previous guides, *Ocean Currents* should prove popular with both teachers and their students. The authors and all the staff of the GEMS and MARE programs are to be commended on this present product. I have read the entire guide and studied all the illustrations and activities, and am impressed with the accuracy of the text and the clarity with which complex concepts are presented. In designing the various hands-on activities for student participation, the authors have shown considerable ingenuity and imagination.

Teachers who have the opportunity to use *Ocean Currents* with their classes are very fortunate, for many—if not most—teachers will not have been exposed to the details of dynamic ocean circulation. Using the guide will be a learning experience for them as well as their students.

Their students will also be fortunate, for it is possible that during their lifetime ocean circulation will be of major concern to people all over the world. Some basic knowledge of the processes involved will help our future adult citizens make rational decisions and choices in the face of potentially major changes in the Earth's environment.

Evidence from several sources suggests that the major currents of the world ocean—both wind-driven surface currents and deep currents driven by density differences—have been much the same for the past 3 million years, ever since North America and South America were joined in the Isthmus of Panama, separating the Atlantic and Pacific. Within that general framework, however, there have been many variations, primarily due to slow changes in the Earth's climate that periodically resulted in ice ages on the continents, which lowered sea level and in turn influenced ocean currents. During warming periods, retreating glaciers on land have at times significantly altered circulation in the oceans. For example, about 16,000 years ago, toward the end of the last ice age, the rapidly melting glaciers covering much of Canada resulted in a vast pulse of cold fresh water entering the North Atlantic. This so upset the circulation that, by about 12,000 years ago, the warm currents flowing toward Europe were stalled and Europe suffered glacial conditions once again.

These natural changes in the past have generally been slow. Today, however, we face a situation unique in the history of our planet. Slow natural changes in our environment are being replaced by relatively rapid changes. These rapid changes are primarily due to human activities that add greenhouse gases and other pollutants to the atmosphere, causing global warming, with far-reaching consequences. A United Nations report issued in January 2001 states that average global temperatures could rise as much as 10 degrees Celsius over the next century. The report concludes that rising temperatures are primarily due to industrial pollution, not changes in the Sun's energy output or other natural causes. Such a rise in temperature will surely cause vast changes in the climate, melting the ice caps on Greenland and in Antarctica, raising sea level, and altering the way water circulates in the ocean.

A recent study by scientists in Japan found that the surface water of the Sea of Japan has risen between 1.5 and 3.0 degrees Celsius in the past 50 years. This warm surface water remains on the surface and does not sink in winter. Thus the vertical circulation is disrupted, oxygen from surface water no longer reaches the depths, and the Sea of Japan, which supports major fisheries, faces a slow death. This phenomenon may well be occurring in many other seas. Global warming is also influencing the water temperature and circulation patterns along the California coast. Intertidal marine animals that in past years were limited to the warmer waters of Southern California are now to be found for the first time in coastal waters of Central and Northern California.

It is entirely possible that later in this century changes in ocean currents will prove to be of as great concern to humanity as other environmental problems are today.

E.C. Haderlie
Distinguished Professor of Oceanography
Naval Postgraduate School
Monterey, California

January 30, 2001

3B

Introduction

Our world is a water planet. Nearly three quarters of Earth's surface is covered by ocean. Looking at a globe from the perspective of the vast Pacific Basin, it appears obvious that Planet Earth should more appropriately be named Planet Ocean! The Southern Hemisphere, with only one third of the land area on earth, could easily be called the Oceanic Hemisphere. The ocean is a major feature that distinguishes our planet from all others in the solar system. The ocean allows life to exist, makes our climate habitable, provides much of our oxygen and food, and transports nutrients, people, and pollution around the globe. It is impossible to understand the biological, geological, or human history of our planet without first understanding something about how the ocean works.

The phrase "sailing the seven seas" is often used in maritime history and literature. In our modern world, however, the consequences of our actions are often global, so it is more important than ever to recognize that there is essentially only one ocean. What is put into one "sea" may very well end up on the beach of another halfway around the world. Looking at a globe from a South Polar perspective, the marine environment can be seen as one interconnected ocean system. The Antarctic continent is surrounded by an "Antarctic Ocean" with three large "extensions," or ocean basins, the Atlantic, Pacific, and Indian basins. Other smaller basins, such as the Arctic and the Mediterranean, can be considered branches of these larger ocean basins. From our land-bound perspective, we may think of the ocean as separating continents, but it also connects and links them together.

Major Content Areas

Ocean currents have been a major force throughout history and have affected living things in many ways. They have determined the course of ships and the migration and settling of people around the globe. Currents affect the climate of the land by moderating temperature extremes. They have a profound influence on the growth of new life in the most productive parts of the ocean.

If you'll pardon a pun, there is a deep current of content running throughout this unit. With the advent of the *National Science Education Standards* and recommendations of the Third International Mathematics and Science Study (TIMSS), there is a recognized need to pursue key topics in greater depth. In this unit, the global topic of ocean currents opens up connections to many other fields, and dovetails with key concepts in physical science.

Ocean Currents delves into major content areas in the Earth Sciences and the Physical Sciences. In addition to numerous aspects of oceanography and environmental science, students gain much direct experience in chemistry and the physical sciences, with special emphasis on the important concept of density. Real-world, language arts, history, geography, and social studies connections abound, as students increase their knowledge and understanding of the role of ocean currents in human migration and trade. They apply their new knowledge to notable historical situations and to creating their own stories featuring ocean currents.

The activities are sequenced to develop the main concepts, and to provide many opportunities for students to make discoveries and connections for themselves. There is some repetition—of a constructive nature—to enable a better grasp and retention of the concepts.

Activity-by-Activity Overview

In **Activity 1**, students are introduced to the vastness of our planet's one, interconnected ocean and the importance of the ocean to all life on Earth. In Session 1, students brainstorm about what they already know about the ocean. In Session 2, they explore a globe with a global exploration worksheet as a guide. In a special "Going Further," students can explore the similarities between a human and our planet.

Activity 2 begins with students sharing their own prior experiences and knowledge about currents. They learn two of the factors that cause currents: wind and water masses of different temperatures. They also learn that currents spread pollutants throughout the ocean. In Session 1, students use a model to make predictions about where are the best and worst places in the ocean to dispose of waste from imaginary countries. They then perform tests, which they interpret, and present their findings. In Session 2, students see how wind sets water in motion.

Only One Ocean *is a companion GEMS guide to this one, also for Grades 5–8, which you may wish to present.* Only One Ocean *focuses on the productive areas of the ocean, the structure and function of organisms of the open ocean (including a squid dissection), and a class investigation of fisheries and the scarcity of ocean resources.*

In **Activity 3**, groups of students examine the relationship of temperature, salinity, and density to currents. They create currents by combining water of different temperature and salinity. They discover the force of the wind and density differences affect current motion at all levels in the ocean. Each group makes a poster describing how what they've discovered relates to real-world currents.

In **Activity 4**, students delve deeper into the concept of density. They layer and examine four liquids in straw cylinders: cold and salty water, cold and fresh water, hot and salty water, and hot and fresh water. In a follow-up discussion, the concept of density at the molecular level is discussed.

In **Activity 5**, students make predictions about whether ice cubes will melt faster in fresh or salt water, then observe a demonstration and hypothesize about the results. The demonstration synthesizes what students have been learning about density-driven currents. Students sum up results and draw conclusions about Activities 1 through 5.

In **Activity 6**, students predict routes for real-life situations, including information on wind-driven surface currents, density- and salinity-driven deep currents, and upwelling and downwelling zones. As they examine stations around the room at their own pace, they draw routes on their data sheet maps with colored pens. Students share their ideas and routes, then show the actual routes on overhead transparencies.

In **Activity 7**, students use world currents maps to make up their own fictional stories involving ocean currents. They are asked to incorporate what they have learned about currents. This serves as an excellent assessment activity and makes a strong language arts connection.

We've included in the book a version of the colorful cover design showing world currents, which you may want to use as a classroom poster.

Drawn from MARE

The activities in *Ocean Currents* are drawn in part from a rich and varied collection of activities developed by the Lawrence Hall of Science MARE program (pronounced as in the Latin for ocean: mär´ a), which stands for the Marine Activities, Resources & Education. Other MARE guides GEMS has published include a companion unit to this guide, *Only One Ocean,* also for Grades 5–8, and *On Sandy Shores* for Grades 2–4.

More About MARE

MARE is an exciting whole-school, interdisciplinary, marine science program for elementary and middle schools. It brings together key science content with activity structures that promote language acquisition by all students. These activity structures are designed to help students talk, write, and draw about their related prior knowledge of a topic, or to distill and summarize what they have recently learned. They have names such as "Thought Swap" and "Think, Pair, Share."

The MARE curriculum also emphasizes "Key Concepts" for each major activity that are not only communicated to the teacher, but directly conveyed to the students.

MARE's yearlong program is highlighted at each school by an Ocean Week or Ocean Month. It engages the entire staffs and student bodies, parents, and communities of hundreds of schools in a comprehensive study of the ocean.

In addition to providing curriculum, MARE offers teacher education inservices and summer institutes based on the most up-to-date scientific and educational research. The program focuses specifically on helping culturally, linguistically, and academically diverse schools to implement high-quality science education that is accessible to all students. Customized, whole-faculty in-services introduce teachers to new methods for developing their own integrated instructional plans based on the MARE curriculum, present marine science content, and help the entire school plan for their Ocean Week. At MARE's two-week, residential Summer Institute, teacher leaders sample hands-on activities, plan schoolwide programs, learn from leading scientists and educators, and participate in exciting field experiences.

MARE's Ocean Week is a whole-school/whole-school-day "immersion" experience that transforms an entire school into a laboratory for the discovery and exploration of the ocean. This intensive educational event creates an exciting atmosphere schoolwide and serves as the centerpiece for yearlong ocean studies. Ocean Week builds a sense of inclusion throughout the school community and improves the general climate and educational culture of the school. Special education, language minority, and mainstream students work side-by-side across grade levels, peer teaching, and tackling special projects. Students have long uninterrupted blocks of time to explore areas of interest in depth. Teachers may receive on-site support from MARE staff, who work at the school every day of Ocean Week, coaching, model teaching, coordinating, and dispensing materials from MARE's extensive multimedia library. Parents are directly involved in the school's academic program.

The MARE curriculum focuses each grade on a different marine habitat and integrates language arts, language development, social studies, and art with science and mathematics. Many key science themes and concepts are explored. Integrating disciplines and linking subject areas, the curriculum helps students understand the overlapping themes of science. Within each curriculum guide, you will find in-depth teacher reference information, hands-on activities, teaching strategies, children's literature connections, planning materials for developing a comprehensive whole-school science program based on the study of the ocean, and suggestions for assembling student portfolios and conducting performance tasks to assess student achievement. Each activity in the MARE curriculum identifies and develops students' related prior knowledge through a rich variety of language experiences, before introducing new topics. Each of the MARE guides can be used, minimally, as a six- to eight-week science unit, or can be expanded and integrated into a comprehensive, yearlong science curriculum covering the disciplines of earth, physical, biological, and environmental sciences.

If you're interested in more information about MARE, please contact:

University of California at Berkeley
MARE
Lawrence Hall of Science # 5200
Berkeley, CA 94720-5200

(510) 642-5008
fax: (510) 642-1055
e-mail: mare_lhs@uclink4.berkeley.edu
Web site: www.lhs.berkeley.edu:80/MARE/

Time Frame

Depending on the age and experience of your students, the length of your class periods, and your teaching style, the time needed for this unit may vary. Try to build flexibility into your schedule so that you can extend the number of class sessions if necessary.

Activity 1: Planet Ocean	
Session 1: Ocean Brainstorm	45 minutes
Session 2: Global Exploration	60–120 minutes or more
Activity 2: Waste Disposal	
Session 1: Student Explorations	60 minutes
Session 2: Demonstrating Wind–Driven Currents	45 minutes
Activity 3: Current Trends—Station Rotations	
Session 1: Introducing the Stations and Making Predictions	60 minutes
Session 2: Completing the Stations	60 minutes
Activity 4: Layering Liquids	
Session 1: Introducing the Challenge	60 minutes
Session 2: Discussing Density	45 minutes
Activity 5: Ice Cubes Demonstration	45–60 minutes
Activity 6: Ocean Routes	60–120 minutes
Activity 7: Message in a Bottle	60 minutes

What You Need for the Whole Unit

Nonconsumables

- ❒ 30–60 ocean-related pictures clipped from magazines or calendars
- ❒ 8–16 (12") inflatable globes showing currents if possible
- ❒ 8 shallow (1½" to 2" deep) clear trays or clear salad containers
- ❒ 17 rocks (about ¼ brick size) or upside down cups
- ❒ 2 large clear deli salad containers or glass baking pans no larger than your overhead can project: about 8" x 8"
- ❒ Pacific Rim Map to make transparency, page 45
- ❒ world map that includes the Pacific Rim (large inflatable globe is best)
- ❒ 1 each of 1A and 1B Station directions, page 63
- ❒ 1 each of 2A and 2B Station directions, page 64
- ❒ 1 each of 3A and 3B Station directions, page 65
- ❒ 4 (12–16 oz.) clear plastic, straight-sided identical water bottles with threaded mouths about 1" in diameter
- ❒ 4 tornado tubes (or Vortex Tubes)
- ❒ 4 dish towels (white preferred)
- ❒ 4 yogurt lids with the rims cut off
- ❒ 4 (6–8 oz.) paper or Styrofoam cups
- ❒ 2 containers—approximately six quart, clear, rectangular, such as a plastic shoe box
- ❒ 4 thermoses with at least 3-cup capacity each
- ❒ 1 sharp knife for slicing potatoes
- ❒ clear measuring cup (1- or 2-cup size)
- ❒ 32 insulated containers—Styrofoam cups with lids or other containers
- ❒ 32 medicine droppers
- ❒ 8 (16 oz.) cottage cheese-style containers
- ❒ 9 cafeteria trays
- ❒ 4 (9 oz.) wide-mouthed, clear plastic cups
- ❒ pouring container with 3–4 cups of water
- ❒ graduated cylinder or other measuring device
- ❒ 2 identical plastic jars
- ❒ 16 overhead transparencies, pages 109–124
- ❒ damp cloth to erase transparency pens
- ❒ posters, charts, and Key Concepts from previous sessions of the *Ocean Currents* unit

Optional Nonconsumables

- ❒ large (24"–36") inflatable globe
- ❒ pump to inflate large globe
- ❒ audiotape of ocean sounds and tape player
- ❒ videotape of ocean scenes and VCR or slides and projector
- ❒ miscellaneous posters showing marine life and ocean scenes
- ❒ world atlas or other geography resources
- ❒ 2 funnels
- ❒ spring or balance scale

Consumables

- ❐ 8–16 Global Exploration student sheets, pages 22–24
- ❐ 8 World Map of Currents data sheets, page 27
- ❐ 32 Waste Disposal data sheets, page 44
- ❐ tap water
  - __ hot water (about 100°F–110°F, which is not quite the hottest available from a home faucet)
  - __ very, icy-cold, refrigerated water
  - __ room temperature water
- ❐ vials of food coloring: 1 each of yellow and green, 2 each of red and blue
- ❐ 24–26 ice cubes
- ❐ 8 Student Explorations with Waste Disposal student sheets, page 43
- ❐ 64 Prediction data sheets, page 66
- ❐ 32 Current Trends data sheets, pages 67–69
- ❐ 1 box of salt (kosher preferred)
- ❐ less than a tablespoon of oil
- ❐ 8 medium-sized, raw potatoes
- ❐ 16 Density Layers Plan student sheets, page 82
- ❐ 16 Density Layers data sheets, page 83
- ❐ station sheets 1–11, pages 125–135
- ❐ student maps, pages 105–108
- ❐ 32 Surface Currents maps, page 109, or Deep Currents maps, page 110

General Supplies

- ❐ about 40 sheets of 27" x 34" chart paper
- ❐ masking tape
- ❐ markers (wide tip, water-based: yellow, black, brown, blue, green, orange, red, blue)
- ❐ overhead projector
- ❐ newspaper
- ❐ 8 sheets of 8½" x 11" white paper
- ❐ about 70 drinking straws
- ❐ 1 sandwich baggie
- ❐ 1 blue permanent marker
- ❐ scissors
- ❐ 2 tablespoons
- ❐ 4 push pins
- ❐ more than 20 marbles
- ❐ 11 stir sticks
- ❐ paper towels
- ❐ transparency pens
- ❐ several sheets of 8½" x 11" paper
- ❐ pens or pencils

Optional General Supplies

- ❐ 6–10 sentence strips or cards

Complete classroom kits for GEMS teacher's guides are available from Sargent-Welch. For further information call 1-800-727-4368 or visit www.sargentwelch.com

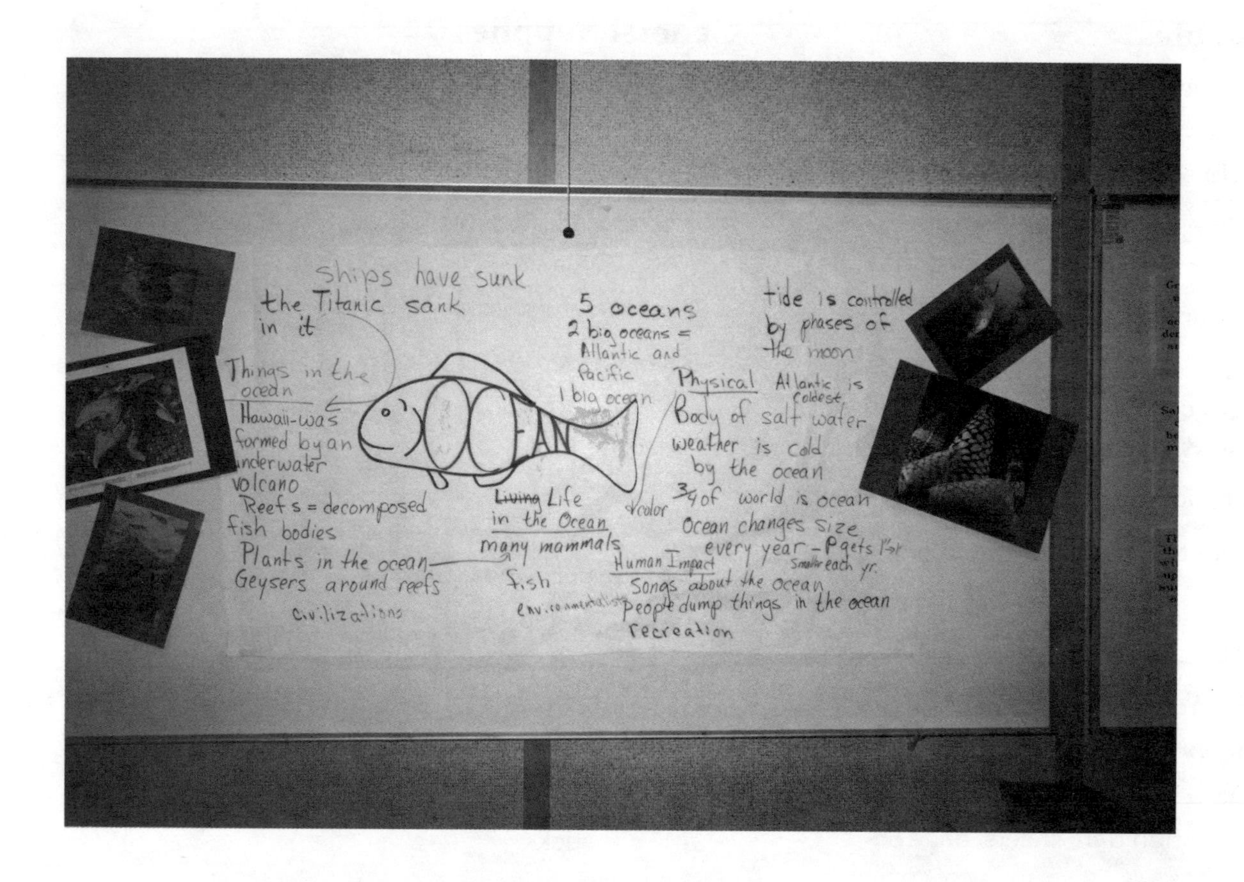
ships have sunk
the Titanic sank in it
Things in the ocean
Hawaii was formed by an underwater volcano
Reefs = decomposed fish bodies
Plants in the ocean
Geysers around reefs
civilizations
OCEAN
5 oceans
2 big oceans = Atlantic and Pacific
1 big ocean
Life in the Ocean
many mammals
fish
Human Impact
Songs about the ocean
People dump things in the ocean
recreation
tide is controlled by phases of the moon
Physical
Atlantic is coldest
Body of salt water
weather is cold by the ocean
3/4 of world is ocean
Ocean changes size every year
color

There's only ONE ocean!
OUR EARTH IS A WATER PLANET COVERED BY ONE INTERCONNECTED WORLD OCEAN THAT CIRCULATES AROUND ALL THE CONTINENTS.
Both need water to exist
Both are mostly water
The Human
H2O
Both mortal
Both are living and moving
Contain salt
portions are similar
Both have living things - organisms
Part of nature
Ciralatory system
Vitamins and minerals
Pollutants
The Planet

Activity 1: Planet Ocean

Overview

Students are introduced to the vastness of our planet's one, interconnected ocean and the importance of the ocean to all life on Earth. Students participate in a wide-ranging brainstorm about what they already know, value, and enjoy about the ocean. They work in teams to explore a globe using a global exploration worksheet as a guide.

How inappropriate to call this planet Earth when it is clearly planet Ocean.
— Arthur C. Clarke

The purposes of this opening session:

- Introduce students to this ocean unit
- Provide a basic orientation to global geography
- Give students a chance to brainstorm together and express what they already know about the ocean
- Provide information on what students know, what misconceptions they may have, and what they want to know more about
- Convey the key concept that there is one, interconnected, world ocean

What You Need

For the class:

❒ 30–60 ocean-related pictures clipped from magazines or calendars
❒ 3–6 sheets of chart paper (approximately 27" x 34") to record brainstorming
❒ masking tape
❒ markers (wide tip, water-based: black, brown, blue, green, orange, red)
❒ sheet of chart paper cut in half lengthwise to display Key Concepts

Optional

❒ large world map
❒ audiotape of ocean sounds and tape player
❒ videotape of ocean scenes and VCR or slides and projector
❒ miscellaneous posters of marine life and ocean scenes
❒ large (24"–36") inflatable globe
❒ pump to inflate large globe
❒ world atlas or other geography resources

The optional audiotape, videotape, slides, posters, and/or maps are especially helpful for students who are English language learners. They create a visual and linguistic context for all the new information presented. The more you have, the more you can use these images to illustrate ideas and concepts. Remember to walk over to an image and point to it or even write on a laminated poster with water-based markers as you say the word. This will help students acquire new vocabulary.

Currents often have several names depending on when in history they were named and the people who named them. The name of a current on a British chart may differ from its name on a chart from the United States or Japan. For example, the current that flows around Antarctica is variously called the Antarctic Circumpolar Current, the Southern Current, or the West Wind Drift.

For each group of four students:

- ❒ 12" inflatable globe (showing currents if possible; two globes would be even better)
- ❒ 1–2 Global Exploration student sheets (long version on pages 22–23 or short version on page 24)
- ❒ 2–4 water-based markers
- ❒ 4 pens or pencils
- ❒ 4–8 sheets of 8½" x 11" paper
- ❒ sheet of 8½" x 11" scratch paper
- ❒ World Map of Currents student sheet (if your inflatable globes do not show currents), page 27

Optional

- ❒ 6–10 sentence strips or cards

The World Map of Currents on page 27 is for the months from May through September. We ask that you use it for purposes of the activities in this unit. Maps of world currents for different months of the year can be found on the MARE Web site: www.lhs.berkeley.edu:80/MARE/.

On the maps you may notice that in the Pacific the Southern Hemisphere gyre and the Northern Hemisphere gyre meet north of the geographical equator. That is because, along with the trade winds, the currents straddle the meteorological equator, which is 5°–8° north of the geographical equator.

Getting Ready

1. Several weeks before beginning this unit, plan your strategy for gathering a large number—at least one for every student in your class—of photographs, pictures, or drawings of ocean images. Magazines and old calendars are good sources. These should include many images relating to the physical ocean, since that is the focus of the unit. These can include watercraft of all kinds and from different periods of history, icebergs, storms, cold weather, hot areas, people on the beach, swimmers, shipwrecks/people lost at sea, waves, currents, wind, rivers, different colors of water, and humans and machines involved in recreation, research, industry, and pollution. **These images are invaluable to prompt student brainstorming, reminding students of what they already know, but they are also valuable as writing and drawing prompts, and to illustrate new ideas and concepts. We recommend that you laminate the images so they can be used many times.**

2. Several weeks before beginning this unit, plan the purchase of the 12" inflatable globes and the posters. A globe showing currents is preferred, but if you can't find one, duplicate for each team a World Map of Currents (page 27). We also recommend getting one large globe, again preferably with ocean currents, for demonstrations and modeling from the front of the room. Regarding posters, many posters of the west coast of North America are fine, but "Marine Mammals of the Gulf of the Farallones" is especially helpful (and inexpensive). If you live on the East Coast or Gulf Coast, you may also want to acquire a poster showing your coastline. See "Resources" on page 145 for more information on these materials.

3. Tape two or three sheets of chart paper (27" x 34") side by side on a wall or chalkboard at the front of the room where you can easily write on them to record the results of the Ocean Brainstorm. Use another strip of paper to write the word "OCEAN" across the top of the charts as a title.

4. Write out the Key Concept for this activity in large, bold letters on a large strip of chart paper and set aside.

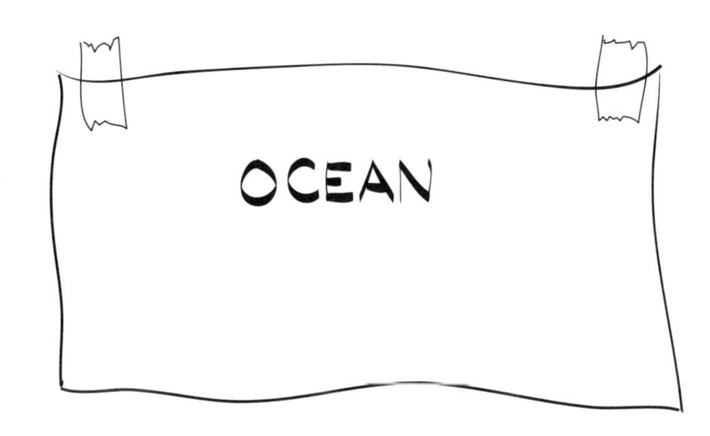

- **There is only ONE ocean!**
  Our Earth is covered by one interconnected world ocean that circulates around all the continents.

  Optional

  Hang the marine life and ocean scene posters, if you have them, around your room where you can reach to draw on them (with the water-based markers and only, of course, if the posters are laminated).

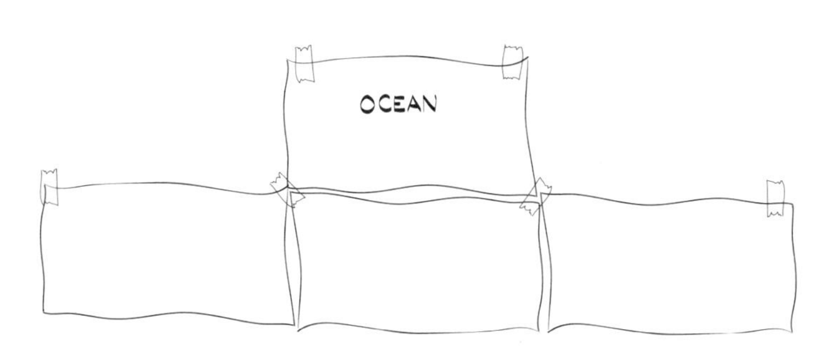

5. For a few days before you begin, ask students to try to notice all the times they see images related to the ocean—on TV, in the newspaper, on billboards, in stores, etc. Ask them to think about their own ocean-related ideas and experiences—what does the ocean mean to them, their family, their culture?

6. Inflate the globes, unless you choose to have the students inflate them for you. Use a pump for the large globe. If you don't have a pump, allow 10–15 minutes to blow it up (sitting down) and several minutes after to catch your breath—it takes a lot of hot air! Display the globe prominently. Keep the globes inflated for the rest of the unit.

You could turn this into a math/science lesson by estimating the volume of the inflated globe, counting the number of breaths it takes to inflate it, estimating lung capacity, etc.

7. Decide which worksheet you will use for Session 2: Global Exploration. The short worksheet (page 24) may be more appropriate for younger students and English language learners. The long worksheet (pages 22–23) requires considerably more written responses and time and includes higher level questions. Duplicate whichever worksheet you decide to use.

Many teachers really liked this brainstorming format. They said it helped build the confidence of their students and allowed them to see their ideas recorded and reflected upon. It was also very helpful in giving students the "prior experience" they needed for the upcoming activities.

Depending on your students experience with brainstorming and familiarity with the ocean and ocean concepts, this activity may need a more or less guided approach. Some teachers said this activity worked better for their students as a whole class discussion with guidance and examples from the teacher.

Giving students the opportunity to discuss what they already know before distributing the ocean pictures puts more of the emphasis on students sharing their own knowledge and experiences rather than focusing solely on what is shown on the pictures. However, for students with very little prior knowledge, you will probably want to pass out the ocean pictures earlier so they can be used for the initial discussion. This gives students ocean images to discuss and a context to assist them in correlating content with new vocabulary as it is presented.

Eavesdropping on students during the brainstorming is important and rewarding. Teachers are often happily surprised by a quiet student, a new immigrant, or a science-phobe who turns out to have fascinating ocean stories or experiences. Publicly recognizing someone as an ocean resource to your class can quickly elevate their status and performance! You'll be amazed at what the magic of the ocean can draw out of your students.

Session 1: Ocean Brainstorm

Ocean Brainstorm

1. Arrange students in groups of four. Show the ocean videotape or slides (optional) if you have them. If you have an ocean sounds audio tape (optional), you can play it in the background throughout the Brainstorm.

2. If you placed poster images of the ocean around the room, point them out. Point to the word "OCEAN" written above your brainstorming charts. Tell the students that each small group is going to think about or "brainstorm" everything they can think of that they already know about the ocean. Give a few examples yourself or call on students to give some examples, such as fish live in the ocean, waves crash, people get food from the sea.

3. Ask students to talk among their group for a couple of minutes about anything they already know about the ocean. They can talk about personal experiences, things they've read or seen on TV—anything that comes to mind.

4. After a few minutes of small group discussions, hold up 2–3 of the pictures of the ocean and tell students that you will distribute a few to each group. **Before you distribute them, explain that the pictures are there to remind students of things they already know, but may not have mentioned in their initial discussion. The pictures may also serve to jog their memories about things they might have forgotten.** And, if they run out of things to talk about, they can always talk about the pictures.

5. Distribute the pictures and again let them talk in their groups for a few minutes. Circulate to check the progress of individual groups, and to listen for interesting information, "experts" you didn't know about, and possible misconceptions.

6. Regain the attention of the entire class. Distribute one pencil or pen and a sheet of scratch paper to each group. Ask them to have one person in their group write down at least five things they already know about the ocean.

7. Demonstrate making a list of ocean facts on the board. Tell them the recorders can write words, sentence fragments, or whole sentences (in any language), or draw images. You might want to use the examples mentioned previously—fish live in the ocean (or just write "fish"), waves crash (or just "waves"), and people get food from the sea.

8. Provide another 3–5 minutes for them to record information from their small group discussions. Tell them you expect each person in the group to contribute at least one thing. Circulate around the room and do some more eavesdropping.

One teacher gave each student a picture and an index card and had them write one thing they already knew about the ocean that was different than everyone else in their group—that way everyone got to write their own instead of only the recorder.

A teacher said she gave students the category headings before the brainstorm because she thought that with her students limited experiences they would only come up with 1–2 categories on their own. Another teacher said that she wrote the category headings on the charts before organizing the information into the cluster diagrams in order to save time.

Organizing the Information

1. Let students know that they're now going to organize what they know about the ocean.

2. Ask for volunteers to share any interesting idea they recorded within their small group. As students volunteer items, create a cluster diagram on the brainstorm charts by arranging similar bulleted items into a few large categories. Typical categories that often emerge are, "The Physical Ocean," "Life in the Ocean," "Human Impact" or "Human Use of the Ocean" and "Ocean Beauty," but only write a heading after you have an item to go under it. **Note:** Be sure to place "The Physical Ocean" at the center of your categories, because later on you will be making connections between that category and all the others.

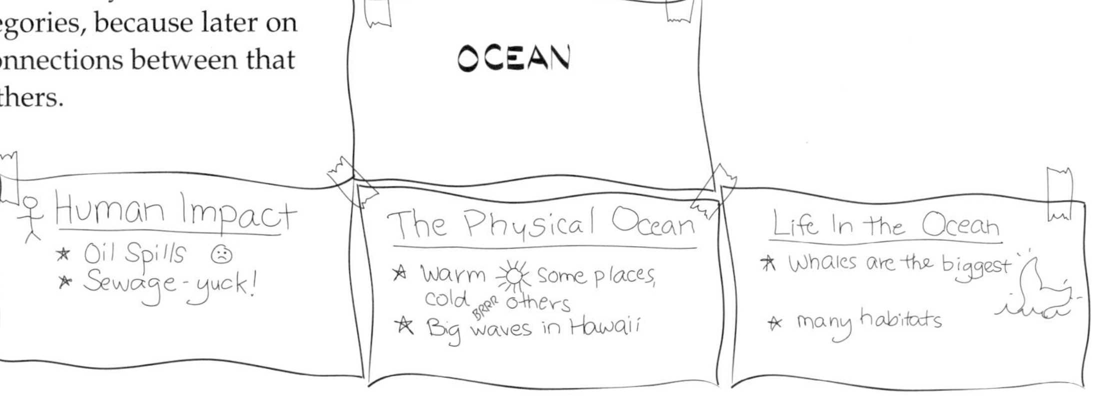

For younger students and English language learners: *Rather than have a recorder write their group list of brainstormed ideas on a piece of paper, you may want to supply cards and/or sentence strips to each group. Students can record key words on the strips and then add them to a "word wall" to build language and literacy skills. Have the students make concept maps from the strips. Revisit the word wall often to add words, work on literacy skills, and refine the concept maps as the students increase their understanding.*

If you had students record information on sentence strips, you might want to bring two groups together to share and combine their strips. Then have them group the strips into categories of similar items (i.e., if they listed fish, sharks, and whales on different strips, they might all be grouped together in one category). Have them decide on a name for each of their categories (for the example above, the heading might be Life in the Ocean). Then have them write their category headings on additional sentence strips and tape them to the walls where everyone can see them. Lead a class discussion of the category headings, noting similarities and differences. If more than one group came up with the same or a similar category heading, discuss how they might merge them. Once the class has agreed on the categories, have each group list their brainstormed items under the category heading most descriptive of their item. Groups can either tape their strips under the appropriate heading or read their item aloud as you record.

3. If a fact could go in more than one category, have students help you decide where to put it. Use arrows between facts and categories to show connections and relationships.

4. When questions emerge ("our group didn't know if . . ." or "we think . . . but don't know for sure") record them on a separate sheet of chart paper labeled "Questions?" As information is gathered on these questions during the activity or unit, you can go back to this chart and write in responses. Unanswered questions can be used to seed future student projects.

5. You may choose to let the brainstorming and clustering go on until the pace at which new ideas emerge slows down, or you may want to end the discussion here.

6. Acknowledge that students already know a great deal about the ocean. **Explain that this unit focuses mainly on the physical ocean.** Explain that "physical ocean," in the context of this guide, means the ocean's motions—especially currents and how they are formed. Use arrows to connect items from the brainstorm to the physical ocean, then ask them if they can think of anything else to add to the physical ocean category, especially about the water and currents.

7. Ask your students to help you relate each of the items in the physical ocean category to those in the other categories. For example, if your physical ocean category says, "the ocean is warm in some places, and cold in other places," you could draw a connecting line to a whale category, and mention that the different water temperatures affect the whales' feeding and migration.

8. Continue making these connections until students realize that everything in the ocean relates to the physical ocean.

9. Tell the class that the charts will be kept posted in a prominent area throughout the unit, so they can refer to them and make additions and corrections. You can also use them to help assess your students' prior knowledge about the ocean and gauge your teaching accordingly.

Dealing With Misconceptions

If misconceptions arise (such as "Sharks kill a lot of people" or "No one hunts whales anymore") be sure to record them in an appropriate category without correcting them. If they are challenged by some students, put a question mark next to them. In the spirit of true brainstorming, this is not the time for editing or negating any individual's ideas. However, make a written note of the misconceptions, and make sure that sometime during this unit students "discover" the correct information. At that point, go back to your chart and make a small ceremony about correcting the class's Collective Body of Knowledge: "Hey, look at this! We used to think that . . . (cross it out), but now we know . . . (write in the new information). No individuals need be "wrong." The group's collective knowledge simply changes as it grows.

Another approach is to make sure that there is a question mark after any statements your students are unsure about or that you recognize as misconceptions. Let your students know that the list is not finalized. Explain that most of the unit's activities will be focusing on currents, so you may not have the opportunity to address all of questions, just those dealing with currents.

If the brainstorming session goes very quickly, and includes many misconceptions, you may need to teach the following activities a bit more explicitly.

Session 2: Global Exploration

Global Exploration

1. Distribute one globe to each team of four and give them a few minutes to freely explore it.

2. Using the globes, review how to read lines of latitude and longitude (see illustration).

3. After a few minutes more of exploring, or when you think they are ready, hand out one or two Global Exploration worksheets (long or short version), paper to record their answers, and a few light-colored (blue works well) water-based markers to each team. Make sure the students take turns holding and exploring the globe and acting as the recorder. Assign them Questions 1–5 and tell them that there will be a class discussion of those questions when everyone finishes. They will then do question 6 followed by a discussion, questions 7–9 with a discussion and finally complete the worksheet with question 10.

4. Introduce the term ***ocean basin*** as a way to describe the Atlantic Ocean, Pacific Ocean, and Indian Ocean areas. Mention that there are other smaller basins (such as the Arctic Ocean and Mediterranean Sea) that are connected to the larger ones.

5. **Remember to wash off the globes after completing question 11 so they are not stained by the markers.**

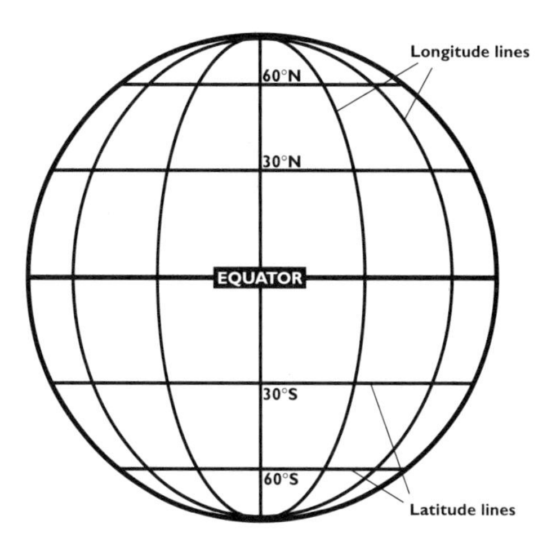

Many teachers said the inflatable globes, and drawing on them, were an "incredible hit with the kids–they almost needed 30 minutes just to explore them because the excitement was so great." Students said "it was mind-boggling—like a two for one—fun and learning." Other teachers said students kept on task and were really motivated to discover the answers to the worksheet by working with the globes themselves. "Initially students weren't sure you could get all the way around the Earth on the ocean—it turned into a discrepant event for them."

When using the longer version of the worksheet, you may decide to give question 1 to two members of the same team and question 2 to the other two members. The questions are similar although one is about land and the other about water. This will allow more time for answering the question in greater detail and the opportunity for more sharing of information about potential itineraries and experiences.

Whenever possible, refer to and integrate into the class discussion information that students identified in the brainstorm. They will appreciate seeing that what they know is relevant to the new information you are presenting. The large inflatable globe is a wonderful and dramatic teaching tool that can help illustrate points in the discussion. Use it as a prop, pointing out places and features as you refer to them. You can also draw on it with water-based markers (just wipe it clean with a damp cloth within a few hours or it will stain).

Discussing the Global Exploration Worksheet

1. After each team has completed Questions 1–5 of their worksheet, help the class analyze it through the following discussion. **Note:** These questions are from the long version worksheet, and include discussion points to emphasize.

2. Use the large and small globes, water-based pens and world map as you deem necessary.

Questions 1–5

1. See how far around the globe you can draw a line without lifting the pen, always staying on **land**. Describe your longest route (the furthest you can go before lifting the pen) including the continents you traversed. Where did you start and where did you end? How many lines of longitude did you cross? How many lines of latitude did you cross? What mode of transportation (no air transport allowed) would you use, how long do you think it would take, what difficulties might you encounter (political, economic, environmental) and how might you overcome them? What languages and opportunities for exciting adventures do you think you will encounter along the way? Do you think your land route is actually feasible and how will you convince people that it will succeed? [From the tip of South Africa to the Chukchi Sea in Siberia is the longest route.]

Teachers said this session really helped their students understand that oceans do not have boundaries. One said, "the concept of one ocean became very powerful when we were able to trace around the whole globe" and "I saw the little light bulbs going off above their heads—they got the big AHA!"

2. See how far around the globe you can draw a line without lifting the pen, always staying on **water.** Record in words the longest route you could find including a description of where you started, which bodies of water you went through, and which continents you passed by. How many lines of longitude did you cross? How many lines of latitude did you cross? What mode of transportation (no air transport allowed) would you use, how long do you think it would take, and what difficulties might you encounter (political, economic, environmental)? Where do you think the most environmental difficulties would be? Do you think your route is actually feasible and how will you convince people that you will succeed?

You may find that after question 2 is a good point to pause and listen to students' routes and travel itineraries before continuing with the worksheet. The discussion and plans can be very rich as students compare itineraries and expectations while using the globe as an integral part and model for their presentation.

3. Find a view of Earth that shows mostly land. Which continent(s) does it include? [Asia, Africa, Europe.] In which hemisphere is it (mostly) located? Sketch a map of the world from this point of view.

4. Find a view of Earth that shows mostly ocean. Which ocean basins does it include? [Southern and Pacific.] In which hemisphere is it (mostly) located? Sketch a map of the world from this point of view.

5. Have your partner toss you the globe. As you catch it, note where the tip of your right index finger lands. Record if that location is on land or water and the latitude and longitude. Repeat this procedure nine more times. What is your estimated percentage of water to land? Now convert the percentage to a fraction. Do you think this is a valid way of determining the percentage? How might you increase the accuracy? **Note:** Have each group combine their results for a larger sample. It should come out to 70%–75% water. Describe another way that you might have figured out the percentage of water to land? [One possible answer: use blue and brown Post-it® notes, cover water with blue and land with brown and then count the number of Post-it® notes of each color.]

Be sure to emphasize these points:

- Water covers much more surface of Earth than land (about three quarters).
- The Southern Hemisphere is about 80% water and the Northern Hemisphere is about 60% water.
- Most classroom maps in the United States show the United States in the center of the map, with the equator below the United States, and the largest feature on the planet—the Pacific Ocean—split in half and just fringing the sides of the map. This tends to make us think the surface of our planet is mostly land rather than water.
- Many maps give the visual impression that our planet is mostly land with some water in between.
- Explain the several meanings of the word "earth" and "Earth." Earth (capitalized) is the name of our planet and consists of everything from the center to the top of the atmosphere; earth is the soil, dirt, and ground on the Earth.
- The ocean is the most dominant, most important feature on the surface of Earth, and helps distinguish our planet from the other planets in the solar system. The ocean does not dominate Earth as a whole since it is very thin (about four kilometers deep compared to the approximately 100,000 kilometer depth of the planet).

Global View: Mostly Land

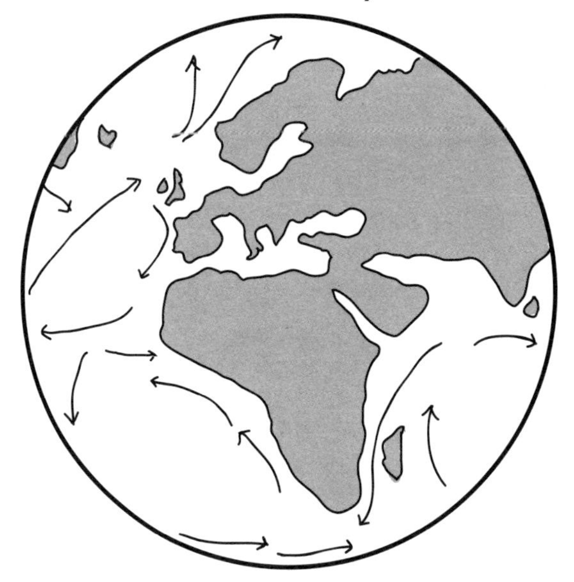

Global View: Mostly Ocean

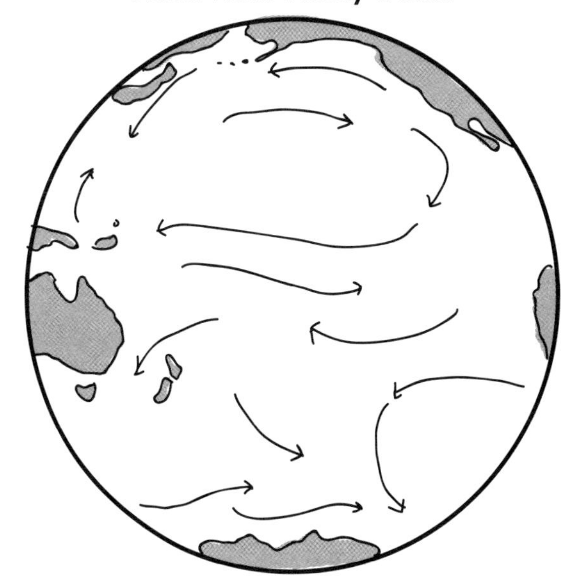

Question 6

6. Can you find any countries that don't touch an ocean, sea, or major waterway? Name five. You probably noticed that most countries do touch or are connected to major bodies of water. What are two reasons that it might it be important for a country to be connected to a body of water? Why do you think landlocked countries have a harder time becoming a world power?

Be sure to emphasize these points:

- Much of the world's food supply comes from the ocean.
- Coastlines have always been places where ideas and goods have been exchanged.
- A huge part of human history, migrations, colonization, trade routes, and wars, up until the time of air travel, is the story of how people have used wind and currents in their travels from one continent to another.

Questions 7–9

7. Find these ocean basins: Pacific, Atlantic, Indian, Southern (or Antarctic), Arctic. In the past it has often been said that there are "seven seas" on Earth. (The word "sea" was then often used for "ocean.") Which additional two basins would you list besides the five listed here? [Answers will vary, but the next two largest basins are the Caribbean and the Mediterranean.]

8. Can you find the boundary where the Pacific Ocean basin ends and the Atlantic Ocean basin begins? How did you decide where the boundary would be? Describe the physical or biological features you used or the latitude and longitude of where you located the boundary. Do you think you could actually see the boundary when crossing over it?

9. Can you find the boundaries of the Southern or Antarctic Ocean basin? What did you use as a boundary line? Do you think you could actually see the boundary when crossing over it? What do you think it would look or feel like?

Be sure to emphasize these points:

- The view of the Southern Ocean shows that the Southern Hemisphere doesn't have much land on it, and **all the oceans are connected.** Antarctica is slightly larger than Australia and much smaller than other continents. On flat maps its size is exaggerated, but not on globes.
- **THERE IS ONLY ONE OCEAN!** Although we call it different names in different places, there are no real ocean boundaries.

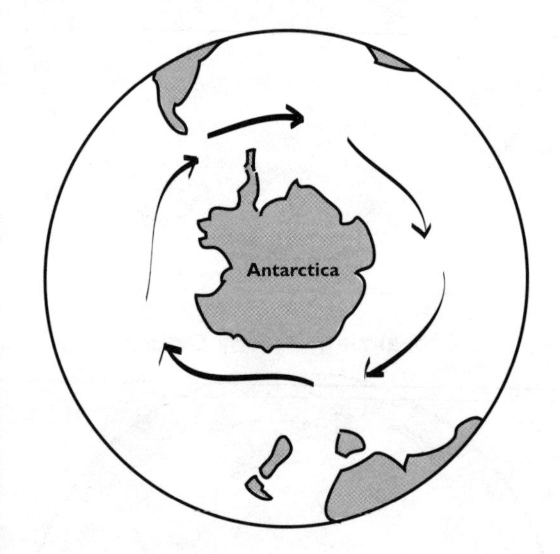

To emphasize that ***there is only one ocean****, you may want to explain that the ocean is one huge, interconnected body of water that surrounds and connects every continent, every island, every plant and animal species, and every culture of the world. There is only one ocean, and it is one huge resource that all people share and depend on. There is no such thing as America's ocean or Japan's or Mexico's ocean—we've only got one and all parts of it belong to all of us.*

Question 10

10. The arrows on the globes represent currents. Find two currents, mark them with the pen, write the name and continent they are adjacent to, and direction they appear to be going. Why do you think it might be important to study about currents even if you are not an explorer?

Be sure to emphasize these points:

- The red arrows indicate warm water currents, and blue arrows indicate cool water currents. As they show their currents, point out whether they are cool or warm and where they come from.
- Ocean currents are huge amounts of sea water moving fairly steadily in a fairly constant direction over long distances.
- Ocean currents move water throughout the world ocean.

You may want to continue the discussion by asking your students why the ocean is important, and why it is important to learn about. Answers will vary, but be sure these points are mentioned:

- *Without an ocean, the surface of our planet would freeze at night and be too hot for most life to exist during the day. The ocean serves as a huge heat sink and it takes a lot of time and energy for it to gain or lose heat.*
- *Take a deep breath. About half the oxygen we breathe comes from plantlike organisms in the ocean.*
- *Just about every culture is connected to the ocean in some way. Much of our language, art, literature, and even religion comes from our marine-related heritage.*
- *The whole world is a marine environment as the ocean strongly affects wind, rain, snow, and air temperature over the entire surface of Earth.*
- *The ocean helps create conditions needed by almost all life to exist on Earth.*

Concluding the Discussion

1. Tell your students that you'll be focusing on ocean currents for the next few sessions.

2. You may want to add that although looking at these globes covered with current arrows may seem confusing, they'll have a much better understanding of what currents are, what causes them, and what some of their patterns are by the end of the unit.

3. In closing the activity, hold up the Key Concept, and have one or more students read it aloud. Briefly discuss how this statement reviews the important ideas from today's activity. Post the concept on the wall near the posters for students to revisit during the rest of the unit.

- **There is only ONE ocean! Our Earth is covered by one interconnected world ocean that circulates around all the continents.**

Reminder: Wash off the globes if you haven't already so they are not stained by the marking pens.

Going Further

Keeping Track of Questions

Revisit your original brainstorming charts and lists of questions. Make any modifications necessary based on the information you have discovered. Questions that are still unanswered can be used for library research topics. You can refer to the charts and lists throughout the unit, including during the "Think, Pair, Share" activity in Activity 6.

Many teachers found this Going Further and its introduction to currents and salinity to be extremely valuable as a lead-in to the rest of the guide. Others thought it was especially effective to place at the end of the guide as a lead-in to a unit on the human body.

A Human Versus the Planet Poster

1. Draw the poster shown here, A Human Versus the Planet, on chart paper and post where all students can view it. Tell students our planet is so big it is hard to imagine how all the parts work together. Sometimes it helps to make a comparison to something smaller that we are more familiar with—so, let's compare the planet to an organism we know: a human, like us!

2. Have students meet back in their brainstorming groups and discuss any similarities they can think of between the planet and a human. Provide 2–3 minutes for the discussion.

3. Ask each group to share any interesting similarities they discussed. Record them on the poster.

4. Now pass out the A Human (page 25) and The Planet (page 26) student sheets. In each group give half the students one sheet, and the other half the other sheet. Tell students to read their sheet, and again discuss in their group any new similarities they can find between the planet and a human.

5. Ask each group to share what they discussed. Record new information on the poster. As students raise the following points you can record them on the poster as shown in the drawing, and provide brief elaboration [in brackets] as appropriate.

- The planet's surface is covered mainly by salty ocean water, and the human body is mostly made up of salty water: blood, sweat, urine, tears. [We are made up of the same thing as the planet.]
- There are salts in our blood and salts in the ocean, but the percentages of each are different for blood and ocean. Also, the total percentage is less for blood than ocean (blood tastes less salty.) [Draw arrows from the words "mostly covered by ocean" to "mostly salt water." There are similarities between the water and blood in our bodies

and seawater. Water can be transferred between us and the ocean. There is only a limited amount of water on the planet and it has been recycling for millions of years. It doesn't matter where the water currently resides. The water in our drinking fountain was once dinosaur blood and will soon return to the ocean—it links us all together as parts of Planet Ocean.]

- The ocean is like our circulatory system or bloodstream. [Both carry food and energy to all parts and carry away waste. Draw a red circulatory system onto the human in your poster, from the heart to all extremities. It doesn't matter what you put in—glucose, medicine, or heroin; or where you put it in—through any vein or artery. It is quickly distributed through the system. We can feed our body quickly with intravenous glucose or we can poison it just as fast with drugs. The ocean works in a similar way although it is much bigger and its circulation is relatively slow. It generally takes decades to millennia for things to get well mixed throughout the ocean. Whatever anyone dumps in the ocean, whether it is sewage or oil from a wrecked tanker, will be distributed through the ocean.]
- Currents in the ocean are the equivalent of flowing blood in our arteries. [Currents globally move ocean water from top to bottom and from hemisphere to hemisphere. Draw a clockwise gyre of currents in the Northern Hemisphere and a counterclockwise gyre in the Southern Hemisphere on the planet in your poster. Show on the inflatable globe by tracing with a marker how a toxic chemical or a piece of trash can enter the ocean off Japan, travel in the North Pacific gyre north and east across the ocean to North America, and down the west coast to Mexico. The pollutant can even cross the equator. It can then get shunted into the South Pacific gyre and end up in Antarctica, South America, or Africa.]
- Ocean pollution can be like toxins in our body. [In many areas, such as the San Francisco Bay, it is recommended that people not eat seafood caught there more than occasionally because of the danger of toxins. There is only one ocean, and when it becomes toxic, so do many ocean organisms we might eat. We can acquire toxins from eating these organisms—and both the organisms and people often concentrate the toxins at higher levels than found in the water itself.]

Some teachers had their students do a Venn diagram using the information on the student sheets.

The goal here is to create a compelling analogy that helps students understand the importance of the ocean—not to make an exact comparison between a human and our planet. Be clear that when we compare water in our bodies to water on the planet, we are comparing mass on one hand to surface area on the other—mathematically problematic and potentially misleading, but still useful. Also, it is difficult to find a precise measure of what percentage of a human body is water. The number depends on whether you include chemically bound (or metabolic) water and other variables. The point is that our bodies are in many ways similar to the planet—and we depend on the health of the ocean very directly and literally to stay healthy ourselves.

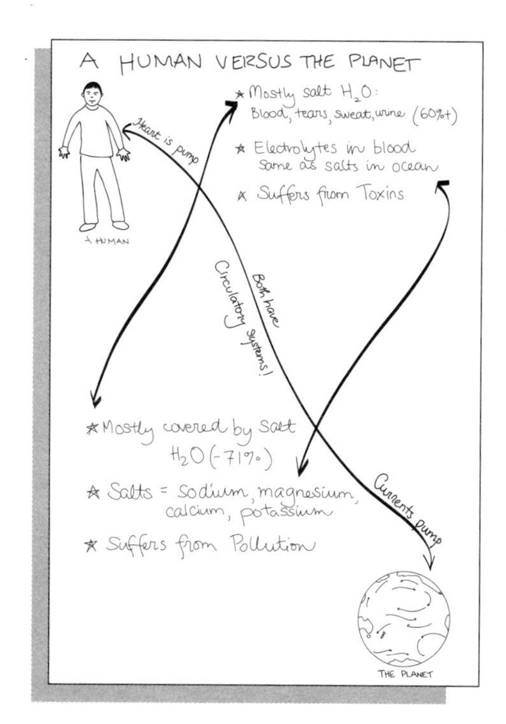

Global Exploration Worksheet (Long Version)

1. See how far around the globe you can draw a line without lifting the pen, always staying on land. Now describe your longest route (the furthest you can go before lifting the pen) including where you started, where you ended, and the continents or countries you traversed.

- How many lines of longitude did you cross?
- How many lines of latitude did you cross?

Use a separate sheet to write about your route as if you were really embarking on a trip. Include the following information:

- What mode of transportation (no air transport allowed) would you use?
- How long do you think it would take?
- What difficulties might you encounter (economic, political, and environmental) and how will you overcome those problems?
- What languages and opportunities for exciting adventures do you think you will encounter along the way?
- Do you think your route is actually feasible? How will you convince people financing you that you have a good chance of succeeding?

2. See how far around the globe you can draw a line without lifting the pen, always staying on water. Now, record in words the longest route you could find including a description of where you started, which bodies of water you went through, and which continents you passed by. Estimate the distance of the route.

- How many lines of longitude did you cross? How many lines of latitude did you cross?

Use a separate sheet to write about your route as if you were really embarking on a trip. Include the following information:

- What mode of transportation (no air transport allowed) would you use?
- How long do you think it would take?
- What difficulties might you encounter (economic, political, and environmental) and where are the most hazards?
- How will you overcome these difficulties?
- What opportunities for exciting adventures do you think you will encounter along the way?
- Do you think your route is actually feasible? How will you convince people financing you that you have a good chance of succeeding?

Global Exploration Worksheet (Long Version)

3. Find a view of Earth that shows mostly land. Which continent(s) does it include?

- In which hemisphere is it (mostly) located?
- Sketch a map of the world from this point of view and attach it to this sheet.

4. Find a view of Earth that shows mostly ocean. Which ocean basins does it include?

- In which hemisphere is it (mostly) located?
- Sketch a map of the world from this point of view and attach it to this sheet.

5. Have your partner toss you the globe. As you catch it, note where the tip of your right index finger lands. Record if that location is on land or water and the latitude and longitude. Repeat this procedure nine more times, for a total of 10. Record your results.

- What is your estimated percentage of water to land? Convert the percentage to a fraction. Do you think this is a valid way of determining the percentage? Why or why not?
- How might you increase the accuracy?
- Describe another way you might have figured out the percentage of water to land?

6. Name five countries that don't touch the ocean or a major waterway.

- You probably noticed that most countries do touch or are connected to major bodies of water. What are two reasons that it might it be important for a country to be connected to a body of water?
- Why do you think landlocked countries have a harder time becoming a world power?

7. Find these ocean basins: Pacific, Atlantic, Indian, Southern (or Antarctic), Arctic. In the past it has often been said there are "seven seas" on Earth. (The word "sea" was then often used for "ocean.") Which additional two ocean basins would you list?

8. Can you find the boundary where the Pacific Ocean basin ends and the Atlantic Ocean basin begins? How did you decide where the boundary would be? Describe the physical or biological features you used or the latitude and longitude of where you located the boundary. Do you think you could actually see the boundary when crossing over it?

9. Can you find the boundaries of the Southern or Antarctic Ocean basin? What did you use as a boundary line? Do you think you could actually see the boundary when crossing over it? What do you think it would look or feel like?

10. The arrows on the globes represent currents. Find two currents, mark them with the pen, and write the name, continent they are adjacent to, and direction they appear to be going. Why do you think it might be important to study about currents even if you are not an explorer?

Global Exploration Worksheet (Short Version)

1. See how far around the globe you can draw a line without lifting the pen, always staying on land. (This is the furthest you can go before lifting the pen.) Describe your longest route, including where you started, where you ended, and the continents or countries you traversed.

2. See how far around the globe you can draw a line without lifting the pen, always staying on water. Describe the longest route you could find, including where you started, which bodies of water you went through, and which continents you passed by. Estimate how long the route was in kilometers or miles.

3. Find a view of Earth that shows mostly land. Which continents does it include and in which hemisphere is it mostly located?

4. Find a view of Earth that shows mostly ocean. Which ocean basins does it include and in which hemisphere is it mostly located?

5. Estimate in fractions how much of Earth is covered by water and how much is covered by land. How did you figure this out?

6. Name three countries that don't touch the ocean or a major waterway.

7. Find these ocean basins: Pacific, Atlantic, Indian, Southern (or Antarctic), Arctic. In the past it has often been said that there are "seven seas" on Earth. (The word "sea" was then often used for "ocean.") Which additional two ocean basins would you list?

8. Can you find the boundary where the Pacific Ocean basin ends and the Atlantic Ocean basin begins? How did you decide where the boundary would be?

9. Can you find the boundaries of the Southern or Antarctic Ocean basin? What did you use as a boundary line?

10. The arrows on the globes represent currents. Find two currents, mark them with the pen, and name them.

A Human

- More than 60% of the mass of the human body is made up of water.
- Our circulatory system (bloodstream) carries food and oxygen to all parts of our body, and carries away waste.
- Our heart drives our circulatory system by pumping blood throughout our body.
- Our bodies can tolerate and eliminate large amounts of toxins, but we may become sick or even die if the toxin level becomes too great.
- Our circulatory system quickly distributes medicine, sugars (glucose), drugs, etc., entering our bloodstream from any vein or artery in our body.
- The salinity of human blood is about 1% (10 parts per thousand).
- Most of the water in the human body (in blood, tears, sweat, urine) is salty.
- A pint of blood will have the same composition no matter what part of your body it is drawn from.
- Some of the major salts (or electrolytes) in our blood are made from sodium, chlorine, sulfur, potassium, calcium, and magnesium.
- The water coming into a person's body from liquids, food, and processes within the body which can produce water (such as using fat stored in the body to produce energy) equals the water going out in the urine and other waste, breath through the mouth and nose, and sweat through the skin.

The Planet

- About 71% of the planet's surface is covered by ocean.
- Ocean water is salty.
- The major salts in ocean water are made from sodium, chlorine, sulfur, potassium, calcium, and magnesium.
- Ocean currents are like blood in a circulatory system. They distribute water, food, and energy to all parts of the ocean.
- The ocean produces and circulates over half of all the oxygen in gas form on the planet.
- There is only one ocean. If you throw a plastic bottle off a ship in the ocean near California, it can wash ashore in Japan, Mexico, or Antarctica.
- The amount of water coming into the ocean from precipitation, rivers, groundwater flows, and other sources equals the amount of water leaving the ocean through evaporation. This is only true when the amount of seawater remains the same and not during ice ages when sea level drops, nor during periods of "global warming" when sea levels can rise.
- The circulation of ocean water occurs through currents.
- The salinity of ocean water is about 3.5% (35 parts per thousand).
- The ocean can tolerate and dilute large amounts of pollution and toxins. But when the levels get too high, the animals and plants in the ocean and everything that eats them are harmed. This can make the ocean unable to support life.
- When toxic chemicals are dumped in one part of the ocean, they may be distributed through the ocean.
- Generally, a cup of ocean water from anywhere on the planet is very likely to have the same salts in it, in the same proportions. Exceptions include water sampled from where a river enters the ocean as compared to water taken from the open ocean.

SURFACE CURRENTS (MAY–SEPTEMBER)

E. = East
S. = South
N. = North
W. = West
C. = Current

F. C. = Florida Current
L. C. = Loop Current
S. Eq. C. C. = South Equatorial Counter Current
C. C. = Caribbean Current
N. Eq. C. C. = North Equatorial Counter Current

S. Eq. C. = South Equatorial Current
A. C. C. = Antarctic Circumpolar Current
S.W. Monsoon C. = Southwest Monsoon Current
N. Eq. C. = North Equatorial Current

WORLD MAP OF CURRENTS

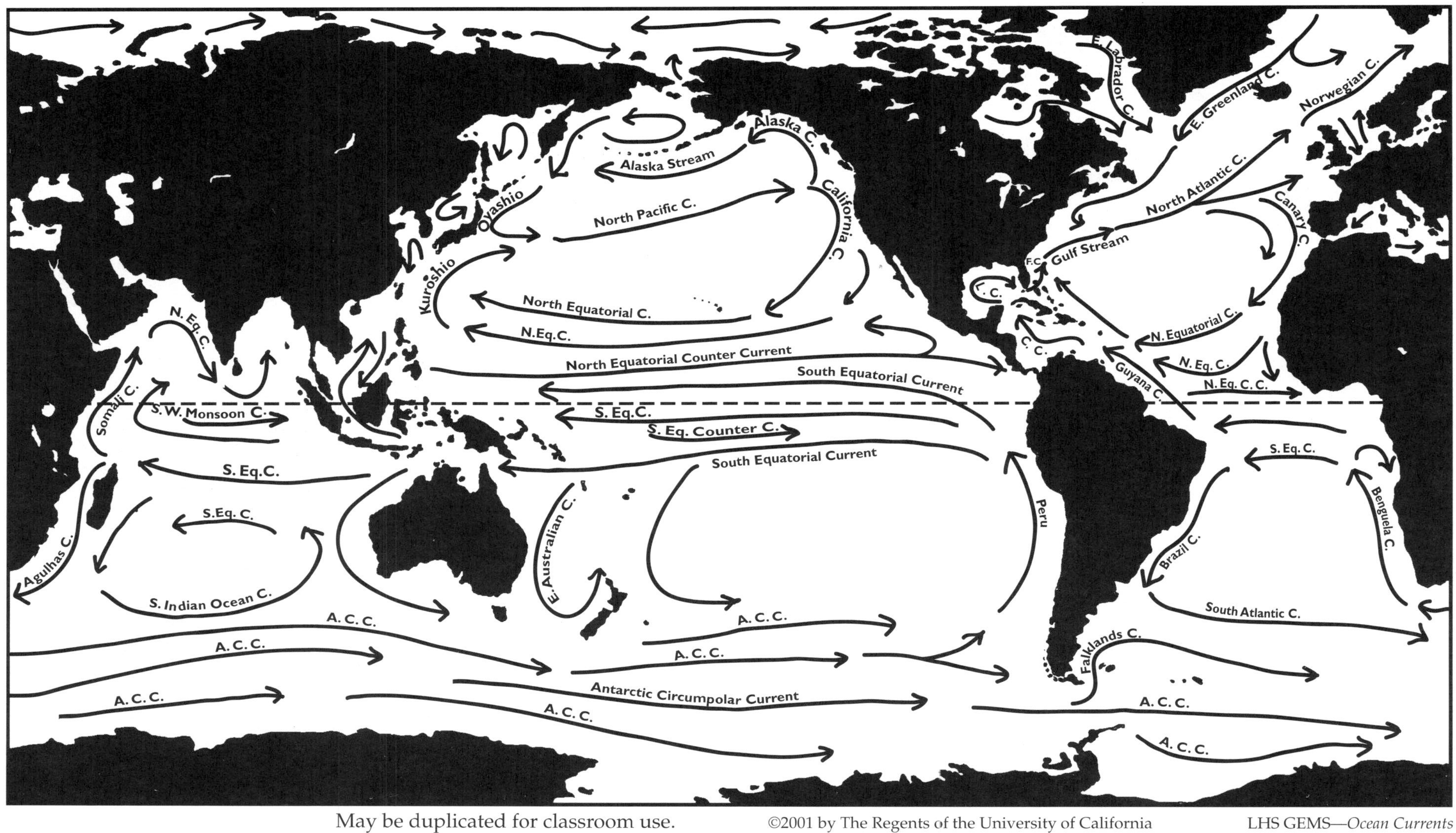

May be duplicated for classroom use.

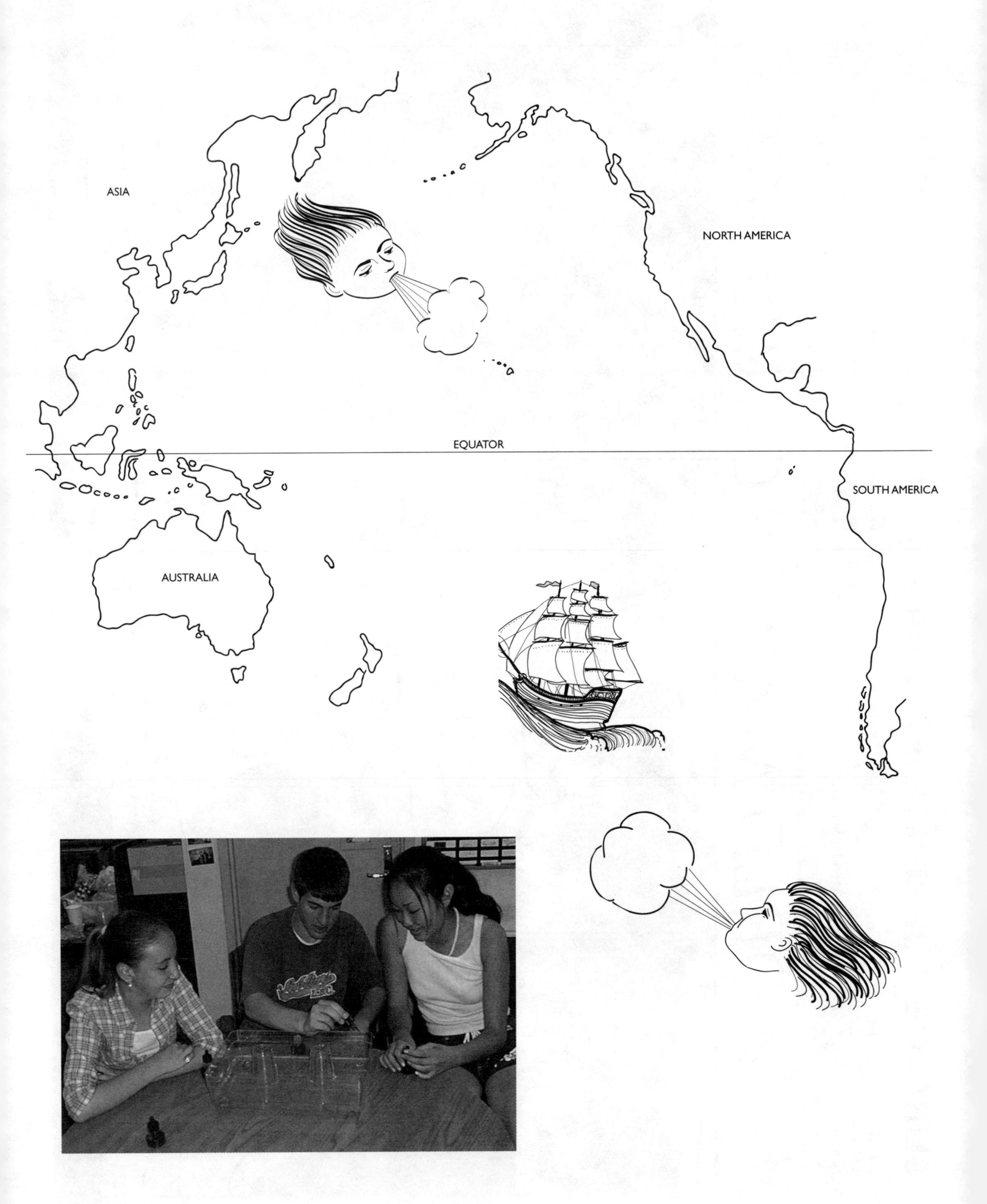
ASIA
NORTH AMERICA
EQUATOR
SOUTH AMERICA
AUSTRALIA

Activity 2: Waste Disposal

Overview

Students begin with the idea of the "Thought Swap" to discuss their personal experiences with currents. They learn that wind and masses of water of different temperatures are two major factors that cause currents. They also learn that currents can spread pollutants, such as toxic waste, sewage, or oil from a wrecked tanker throughout the ocean. Wind sets the surface of the ocean moving, but other forces direct the flow into major gyres circulating clockwise in the Northern Hemisphere and counterclockwise in the Southern Hemisphere.

Student groups make predictions about the best and worst locations in the ocean to dispose of waste from imaginary countries. In Session 1, they perform tests with a simple model of an ocean and continents and interpret and present their findings. In the Wind Driven Currents teacher demonstration in Session 2, an overhead projector is used to model how wind sets water in motion.

The main purposes of this activity:

- Provide students with some initial understandings of two causes of currents—wind and water of different temperatures
- Convey to students that over time currents distribute things dumped at one place into the ocean throughout the ocean
- Introduce the major currents, or gyres, in both hemispheres
- Provide students the opportunity to model the effects of wind-driven currents and gain understanding of how models can be used to represent real-world phenomena

You may also want to discuss the advantages and limitations of the models used in these activities.

Teachers comments:
"Because of the last activity, the kids seemed very eager, they were ready to go—listened carefully to instructions and had a lot of fun with this, asking 'when can we repeat this activity' and 'can we do this again?'"
"They loved the competition between 'countries' and really enjoyed their discoveries. Students generated much concern about the pollution and determined that there is no best place to dump waste in the ocean because it all mixes in the end."

What You Need for Student Explorations

For the class:

❒ overhead projector
❒ sheet of chart paper cut in half lengthwise
❒ enough newspaper to cover the tables
❒ masking tape

For each group of four students:

- ❒ set of 4 colored markers or crayons: yellow, red, green, and blue
- ❒ shallow (about 1½" to 2" deep) clear tray or clear salad container
- ❒ water to fill the tray or container 1" deep
- ❒ white paper or an additional copy of the Waste Disposal data sheet to place under the clear container
- ❒ 2 rocks (about ¼ brick size) or upside down cups
- ❒ 4 vials (¼ fl. oz.) of food coloring (yellow, red, green, and blue)
- ❒ pencil
- ❒ ice cube
- ❒ 4 drinking straws
- ❒ sheet of chart paper
- ❒ small inflatable globe
- ❒ Student Explorations with Waste Disposal student sheet, page 43

For each student:

- ❒ Waste Disposal data sheet, page 44

What You Need for Teacher Demonstration

For the class:

- ❒ 2 large clear deli salad containers or glass baking pans no larger than your overhead can project: about 8" x 8"
- ❒ water to fill pans ¾ full
- ❒ 5 straws for five students to create wind
- ❒ sandwich baggie
- ❒ rock "island" to fit in pan (about the size of ¼ brick) or upside down cup
- ❒ Pacific Rim Map, page 45
- ❒ world map that includes the Pacific Rim (large inflatable globe is best)
- ❒ 2 sheets of chart paper (approximately 27" x 34") or white board or one long piece of butcher paper or chalkboard
- ❒ overhead projector
- ❒ blue permanent marker
- ❒ scissors
- ❒ masking tape
- ❒ sheet of blank paper for each student
- ❒ pencil for each student
- ❒ World Map of Currents from Activity 1, page 27

Getting Ready

For the Student Explorations:

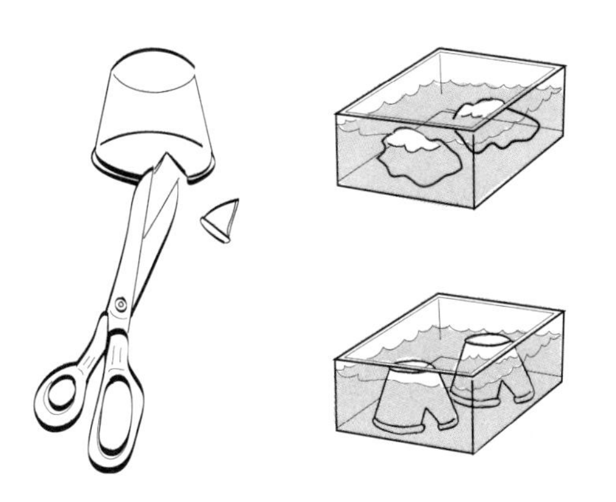

1. Make four copies of the Waste Disposal data sheet, page 44, and one copy of the Student Explorations with Waste Disposal student sheet, page 43. Set up the materials on tables or trays for teams of four students to work together.
    - set of 4 colored crayons
    - clear tray or clear salad container
    - fill the tray or container 1" deep with water
    - 2 rocks (about ¼ brick size) or upside down cups with triangle shape cut out of base
    - 4 vials of yellow, red, green and blue food coloring
    - pencil
    - 4 Waste Disposal data sheets
    - sheet of white paper or an additional copy of the Waste Disposal data sheet on which to place clear container

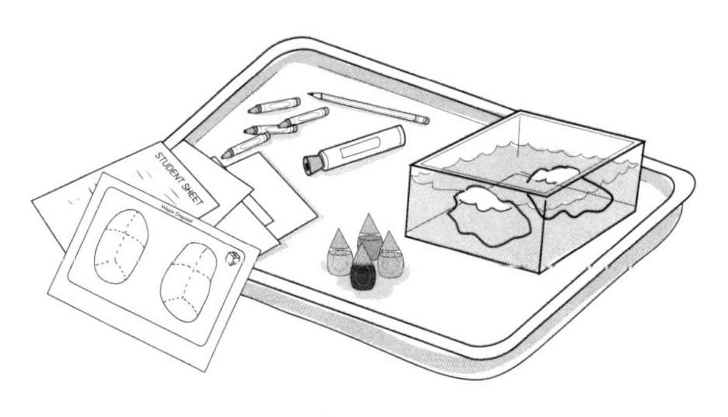

Set aside for each team of four:

- large ice cube (about 1½" x 2½" or several smaller)
- Student Explorations with Waste Disposal student sheet
- drinking straw
- sheet of chart paper
- small inflatable globe

Important Note
If you plan to do this session with classes in rapid succession with only a 5-minute break in between (like many middle schools and science teachers), you will want to set up multiple trays so the clean up and repeated set up can wait until you have more time.

2. Cover the tables with newspaper to catch splashes.

3. Set up a location for students to move their chairs to sit away from the materials as you demonstrate the procedure and for later discussion of the results.

4. Obtain ice cubes and a way to keep them cold until they are needed.

5. For each team of four, mark an "X" on their four copies of the Waste Disposal data sheet to represent their "country." Assign a different country to each team.

Rather than choosing a different country for each team, some teachers let students decide for themselves where they wanted their country on the rock continents. The students then marked the "X" on their own data sheet. One teacher had each team member choose a different country and then they competed among themselves as to which of them could get rid of their own waste without being contaminated by someone else's. Both of these scenarios encourage lively discussions about where to place their "X." All of these ways work well—just keep in mind that it is best if most teams choose different countries.

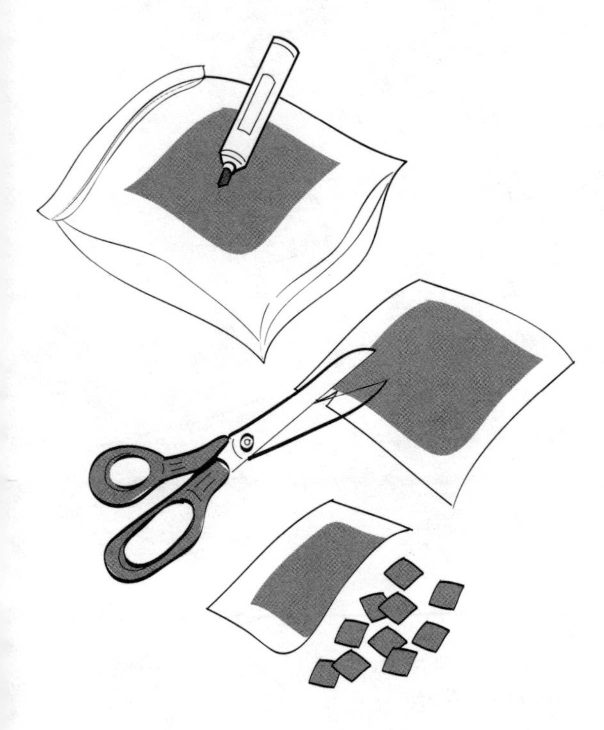

For Teacher Demonstration:

1. Have five straws on hand for students to create wind.

2. To make the plastic confetti, use the colored permanent marker to color an area about 4" x 4" on the sandwich bag plastic. Cut this into quarter-inch squares. You will need about 20 small squares for the demonstration.

3. Make a transparency of the Pacific Rim Map, page 45.

4. Tape the two sheets of chart or butcher paper to the wall (or use a white board). You can use a chalkboard, but the white background is best. If you are using one long sheet of butcher paper, one end will be for the map, the other will be left blank for now. Place the Pacific Rim Map overhead transparency on the projector. Draw the map outline onto one sheet of chart paper or one end of the long sheet of butcher paper by tracing the projected image. **Be sure to draw in the equator, but don't add any of the currents yet.** (The students will be "discovering" the currents for themselves and then later comparing them to the actual currents.) Then remove the map transparency.

5. If you plan on reusing the chart or butcher paper map for another class, laminate it or put contact paper on it. During class you can draw on it with water base markers which can be wiped off with a damp cloth.

6. Place one of your salad containers on the overhead projector and note where it appears on the paper. Make sure the left and right edges of the container line up approximately with Asia and the Americas.

7. Fill the two containers three-quarters full of water.

8. You will want to practice making wind, current gyres, and wind with an obstacle before doing it with the students.

9. Write out the Key Concepts for this activity in large, bold letters on a strip of chart paper and set aside.

- **Things dumped into the ocean may be distributed by currents throughout the ocean.**
- **Wind and the temperature differences between masses of water are two factors that cause currents.**
- **Winds blowing across the surface of the ocean—combined with other factors—cause major circulating currents, or gyres.**

Among the other factors involved in the creation of the gyres is the spinning of the Earth.

Session 1: Student Explorations

Thought Swap

In Thought Swap, students build on their active listening skills by learning how to hold short interesting discussions about currents with a variety of different partners. This activity helps students talk and write about their related prior knowledge. It emphasizes short discussions with different partners, cooperation, and social skills development. It creates opportunities for students to use language in a nonthreatening, but highly relevant setting.

1. Tell students they will get a chance to talk with different classmates. They need to cooperate, follow directions, and talk quietly with each of their partners.

2. Ask students to recall what a good listener should do. [Show interest in what is being said and don't interrupt.] In Thought Swap, both partners will be able to discuss each question or topic. To have a good discussion, each partner should be a good listener and speak clearly when it's her or his turn.

3. Have students stand shoulder to shoulder to form two parallel lines, so each person is facing a partner. Students standing side by side should be at least 6" apart.

4. Tell students you will be asking a question or giving them an idea to talk about with their partner who is facing them. They will have about a minute to talk.

5. Pose the first question for students to discuss from the list which follows step 7 below. Walk along the two lines to help shy or resistant partners get started and listen to their conversations. When you call time, have a few students report something that their partner told them.

You may want to pass out some ocean and water-related pictures to help spur discussion.

6. Before the next question, tell students the line needs to move along. Have *one* of the lines move one position to the left so that everyone is facing a new person; the person at the end of that line walks around to the beginning of the line. Everyone now has a new partner.

7. Repeat steps 5 and 6 until you've asked all these questions:
- When and where was the last time you went swimming or wading?
- Did you feel a current? What is a current and how could you tell that a current was happening?
- Where and when have you ever noticed cold and warm layers of water?
- Describe some of the ways you can think of to make a current in a pool, tub, or even a glass of water.

8. Write the last question on a sheet of chart paper and record their answers. The chart will be referred to at the end of Activity 5.

9. Now is a good opportunity to create a new heterogeneous group of four for the upcoming hands-on activity based on the students' last position in line.

Some teachers were concerned that the Thought Swap activity would take too much time, their students would be too silly, or it was at too low a level for their students—so they decided to skip it. However, many others tried it with their students and told us "This is a GREAT activity—the kids enjoyed it and it's something I will use from now on." (Grade 8). "I was not sure it would work—but it did! It helped them to relate the content to their own experience and it was a good lead-in to currents." (Grade 8). Use your own judgment since you know your students best, but be aware that many teachers not only said they would use this activity again but they planned to adapt it to different subjects.

Introducing the Activity

1. Ask your students if they know where garbage and wastes are dumped. If no one mentions dumping in the ocean, bring it up yourself.

2. Tell them they are going to do an activity that simulates the dumping of wastes at sea, and track its movement.

Teachers said this activity was "rich in information through inquiry." One teacher "enjoyed watching the groups attempting to place the 'waste' where it would not come back to contaminate them."

Explaining the Procedure

1. Explain the following procedure to your students as you demonstrate with your set of materials on the overhead projector.

a. Cover your table with newspapers.

b. Fill your salad container or tray with water about 1" inch deep.

c. Place your continents (rocks or upside down cups) as shown on the Waste Disposal data sheet.

d. Notice on the Waste Disposal data sheet that the location for the ice cube (not distributed yet) is assigned and has already been marked. Every team will place their ice cube in the same location within their container.

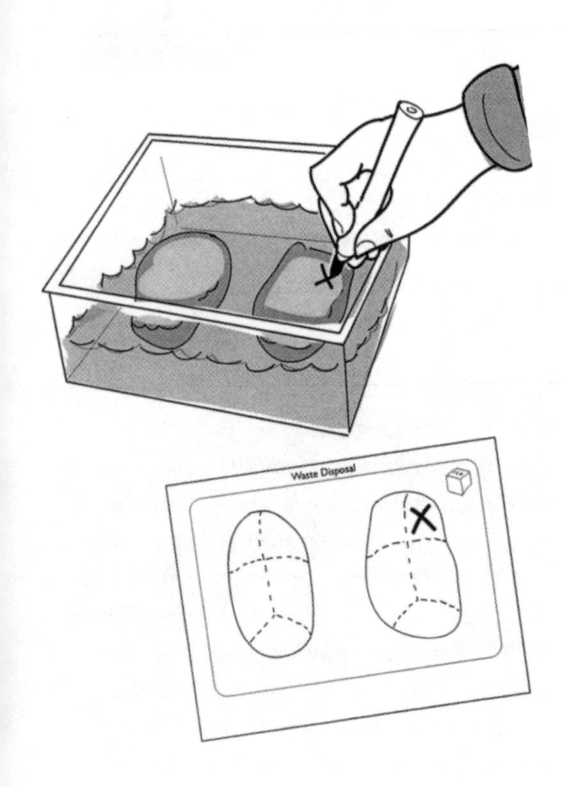

e. Your team has been assigned a "country" marked by an "X" on the Waste Disposal data sheet. On the recording sheet, the dotted lines indicate the boundaries of different countries on the continents. Each team has been assigned a different country that is attempting to dump wastes in the ocean in such a manner that their own country will not be contaminated.

f. Make a mark representing the location of your "country" on one of the two rock "continents." (If you are having student teams decide where their country is located, have them make an "X" on the data sheet to mark the country they select.)

Optional

Name your country.

2. Tell your students their imaginary country has decided to dispose of waste in the ocean. Their team's job is to find the best and worst locations to dump it.

a. *Best location:* Waste will not spread to other parts of the ocean.

b. *Worst location:* Waste will spread most quickly to other parts of the ocean.

A few teachers said they used food coloring (over their carpets) with great trepidation. Fortunately, they were very pleasantly surprised. One said, "It turned out fine! Very messy, (thank goodness—our carpet was spared!) but enjoyable and worth it." Nonetheless, many teachers recommended that control of the food coloring be in the hands of the most reliable student in each group!

3. Tell students that as a team they will decide the location of their four test sites in the ocean. They will use a different color of food coloring for each site.

4. Tell students they will simultaneously place a different color of food coloring at each of the four sites. Each student will use **four drops** of food coloring—**except the yellow site—where eight drops should be placed.** Each color will be monitored by the student who placed it, and who will record its progress on her own Waste Disposal data sheet with a matching colored crayon or marker. Each student will also record the direction of any currents that form by making arrows with a pencil on their sheet.

Some teachers preferred to have students add the food coloring one color at a time and record the progress of their assigned color on ***one*** *data sheet. All the results are colored on the same sheet rather than each member having their own sheet—that way they also see the mixing on paper.*

5. Demonstrate dropping the four different colors of food coloring at each of the sites you have chosen. Ask students to describe what they observe. Then ask them what is missing from your demonstration setup that they will be using when they do the activity. [Ice cubes.] Tell them that you are modeling doing a ***control*** for this activity. Discuss what that means and why it is important to have a control when doing experiments. (Don't tell the students yet, but in this case, the control is important so students will later understand that, in part, it is the cold ice melting off the ice cube which sets the water in motion—the current—which then spreads the food coloring "waste.")

If you have time, you might have the students actually do the controlled exploration themselves—without the ice cubes or wind.

6. Let them know there will be no wind at the beginning, but there will be wind later on. Explain that the wind direction is marked on their maps and they may want to keep it in mind when choosing their test sites.

7. Explain that when they have set up their container on top of the white paper, placed their rocks, indicated their "country" with an "X," chosen their disposal sites, and decided who will monitor each disposal site, you will then distribute ice cubes and food coloring. **Remind them to be very careful not to jiggle or blow on the tray once they have added the food coloring.**

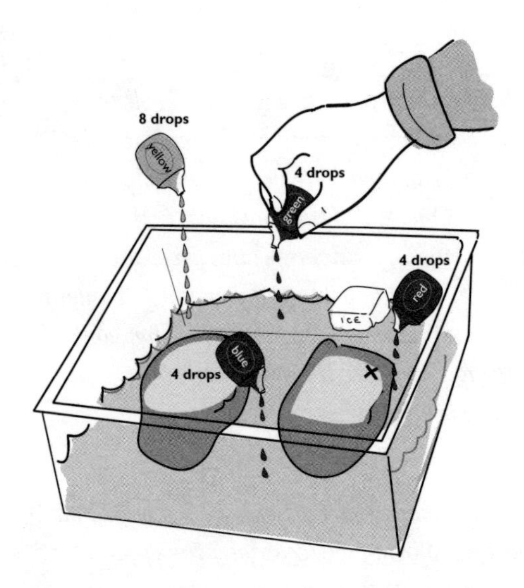

One teacher said "The students were shocked to find that wherever they dropped their pollutants, it affected the whole ocean. Very eye opening."

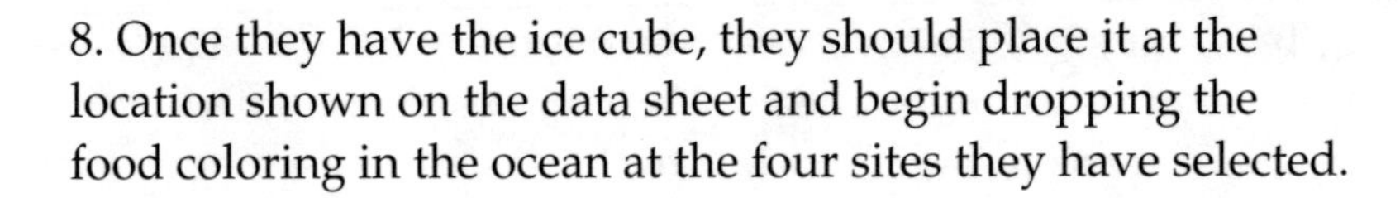

8. Once they have the ice cube, they should place it at the location shown on the data sheet and begin dropping the food coloring in the ocean at the four sites they have selected.

Placing and Tracking Waste in the Ocean

1. Ask students if they have any questions about what they are going to do.

2. Circulate as student teams set up the container, place the rocks (the continents), indicate their country, select the disposal sites, and decide which student will monitor what happens at which site.

3. Distribute the food coloring and ice cubes and have the students start their explorations.

4. Circulate among the groups, reminding them as needed of the procedure.

5. While the students are working, draw a quick sketch of the Waste Disposal data sheet on the board to use for the whole group discussion.

6. Have each team combine the results from all four of their sites onto a piece of chart paper. They should each draw in the colors and current flow they were responsible for monitoring to this one large sheet. This poster can then be used to describe the results of *their* exploration to the class. Also have them record an "X" to mark their worst location, and an "O" to mark their best.

7. Distribute one copy of the Student Explorations with Waste Disposal student sheet to each team. Have them choose a recorder to write down at least four observations they made during their explorations (one of the observations should compare the results of the control with their results.) Then based on these observations, they should generate questions about what they observed. The last part of the worksheet asks them to discuss possible generalizations they could contribute to a class discussion. Each team member should contribute at least one observation and one question and together they can discuss possible generalizations.

Discussing the Results

1. When you are ready (after 15–20 minutes), tell each team to record on the Waste Disposal data sheet copied on the board an "X" to mark their worst location, and an "O" to mark their best. The materials should be left in place, since students will soon return to them to add wind.

2. Gather your students **away from the materials**. Ask each team to share something they observed using their large chart to illustrate their results. [They should have noticed that the food coloring moved around the water in the tray at different rates. Some sank and moved along the bottom, and some moved to the surface. Some colors may have formed barriers between them, or had one slide along on top of the other. Any food coloring near the ice cube will tend to flow away from the cube.]

3. Have them share the questions they wondered about. To get them started in the discussion, you might ask, "What made the food coloring move?" "What do you think made it move so fast?" Ask if anyone has ideas about how they might go about finding the answers. Are some of the questions ones they might answer with further investigations? Where might they find the answers to the other questions?

4. Now ask them how well they think this model simulates the real ocean. What are the limitations? Can they think of some ways to make it more representative of the real ocean?

5. Based on their observations and the limitations of this model, what generalizations might they now make? [Currents formed by cold melted water from the ice cubes sinking down and moving along the bottom. In the ocean, currents may also form as cold water sinks down and warmer water rises up.] Ask them how we might further test these generalizations with this model. How might we test our ideas in the ocean?

Ask students how ocean water gets cold. Is it just when the snow melts and flows into the ocean?

6. During this discussion, encourage your students to draw or illustrate their ideas on the map on the board as they describe them to the class. Note where the best locations were (the "O"s)—the ones that spread the least. Ask them why they think these didn't spread as much as the other locations. Point out their worst locations (the "X"s). Ask why these spread so quickly. Ask if the continents affected the currents at all. How?

*This activity allows students to model and visibly track how a pollutant can spread throughout the world ocean. Many toxic pollutants, however, are NOT visible and cannot be visibly tracked. Also, have students note that the yellow "pollutant" is **twice** as concentrated as the others. Point out that in this case the fact that the yellow is not as easily seen does not mean it is less toxic than the other colors.*

If the setup is not disturbed, the food coloring will take at least 45 minutes to mix throughout the model. If the food coloring has completely mixed before introducing the wind, have the students set up the experiment as before, but this time immediately introduce the wind.

Not everything that gets put into the ocean spreads everywhere. Some things sink to the bottom and get buried. Others may be eaten, or decay, or be broken up by other chemical processes.

There is more information about the formation of currents due to wind, temperature, and other factors in "Behind the Scenes" on page 141.

7. If your students bring up the idea that the best location is where it is least likely for the waste to reach *their* country, ask if that's fair to the other continents and countries. Encourage discussion.

Introducing Wind

1. Tell your students they are now going to introduce another factor that contributes to ocean currents—**wind.** Explain that, taking turns, one member of their team at a time will blow "wind" through a straw **only in the direction shown on the map.** Tell them to observe the food coloring as before, but this time **not** to draw their results on the map.

2. Pass out four straws to each team.

3. When most food coloring has been spread everywhere in the trays (fairly quickly), collect the materials, and regain the students' attention. Ask them what happened. [Among other things, they should have noticed that the food coloring spreads everywhere.]

4. Point out that this spreading is similar to what happens in a person's bloodstream when something, such as medicine or drugs, is injected into it.

5. Explain that it can take decades or even millennia for things to get mixed throughout the ocean.

6. You may want to remind students **there is only one ocean** and water of different temperatures and wind are two factors that cause currents which spread things throughout the ocean. Hold up the Key Concept, and have one or more students read it aloud.

- **Things dumped into the ocean may be distributed by currents throughout the ocean.**

Briefly discuss how this statement reviews the important ideas from today's activity. Post the concept on the wall near your posters for students to revisit during the rest of the unit.

Session 2: Demonstrating Wind-Driven Currents

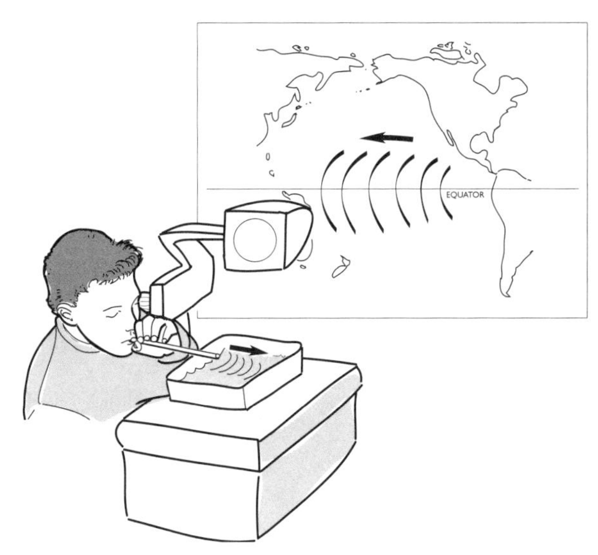

1. Tell the students that you are going to simulate the wind blowing over the surface of the ocean at the equator.

2. You will need five student volunteers to help you. You should have them take turns creating the wind, following your directions carefully. As students blow to create currents, direct them, and also point things out to the rest of the class on the overhead projection. The reason to have five volunteers is so no one student has to blow too much. As soon as a student tires, have another take over.

3. Direct the overhead projector over the Pacific Rim Map which has been traced on the chart paper and taped to the wall or board. Distribute paper and pencil to each student. Have them draw the Pacific Rim outline and equator on the paper.

Some teachers said this works best when you pick "tall winds" (tall students) to simulate the wind blowing over the overhead.

4. Place the first student-with-straw in position to create an east to west (right to left) current along the equator on the projected image. Due to the nature of an overhead projector, the student should be on the reverse side of the container to achieve this.

Oceanographers refer to currents by the direction they are flowing. One reason for this was early sailors needed to know where a current might carry them—toward home, toward rocks. The opposite is true for wind. Meteorologists refer to winds by the direction of their origin. Where a wind comes from tells us a lot about a wind—a northern wind often carries cold air.

5. **Have the student hold the straw nearly level with the surface of the water, keeping the outflow end down low**. The student should hold it far enough back to blow **very lightly** over the surface of the water along the equator. Have each student record what they observe on their Pacific Rim map, and include labels and arrows showing the direction of water and wind movement.

6. After completing their sketches, lead a class discussion about what they observed. [They should notice dark lines made by currents pushed by the wind and describe how the water moves in relation to the wind direction.]

Wind doesn't actually bulldoze the ocean along in exactly the way this activity suggests. The wind-ocean connection is more complicated. A combination of four major factors—the Sun's heat, winds, the rotation of the Earth, and gravity—cause the ocean surface to circulate. The main current gyre flows clockwise in the Northern Hemisphere and counterclockwise in the Southern Hemisphere. *See "Behind the Scenes" on page 141 and "Resources" on page 145 for more information and references about this and other advanced topics.*

7. As needed, you can have a second student take over blowing. Explain that this model simulates how winds blowing over the ocean set the surface of the water flowing in the same direction as the wind. Explain that winds near the equator (the trade winds) generally blow from east to west, causing the east to west Equatorial Current. Draw this current on the Pacific Rim Map on the butcher paper and have students label this current on their own drawings.

Teachers said "the wind demo went great" and "students' enthusiasm for the process was genuine, not to please the system, but pleasure in itself." Another said, "This was a very good demonstration—surprised me!"

Demonstrating Gyres

1. Tell students that if there were no land on Earth, but only ocean, the Equatorial Current might flow east to west all the way around the globe.

2. Explain that the next part of the demonstration will show what happens when the Equatorial Current runs into continents.

3. Drop 10–15 small plastic pieces you cut up previously into the pan and tell the students that by watching them they will be able to see the directions of the surface currents. (Using more than the suggested number of blue pieces starts to bog down the process, making it difficult to see the gyres.)

4. Have the next student with a straw hold the end of the straw down low and level with the side of the pan. Place the student so the wind from the straw will be aimed across the equator at the center of the pan to create a current in the same direction as before.

5. You may have to make slight adjustments in the positioning and blowing of the student(s) to get the Equatorial Current centered enough so the plastic pieces deflect both right and left to form gyres in the Northern Hemisphere and Southern Hemisphere, as shown in the diagram. (Remember, everything you do will be reversed on the overhead, which can make it confusing.)

6. As needed, have another student take over blowing. Have the class again draw what they observe on their own map of the Pacific Rim. Ask the class to describe what they observe. Have a student draw on the diagram on the board the direction in which the colored plastic squares travel, using arrows to illustrate the patterns.

Currents don't stop when the wind stops blowing for a short time. The water has acquired momentum from the wind.

7. If the students don't do so themselves, point out the two large circular currents in the tray, one going clockwise in the Northern Hemisphere, and the other counterclockwise in the Southern Hemisphere. Tell them these currents are called ***gyres.*** Students should add these gyres with labels and arrows to their own drawing if they haven't done so already.

8. Looking at the two gyres, ask your students to predict whether the currents reaching the west coasts of North America and South America would be warm or cold water. [They would be cold, since the gyre in the Northern Hemisphere is moving down from cold northern areas, and the gyre in the Southern Hemisphere is moving up from cold southern areas.]

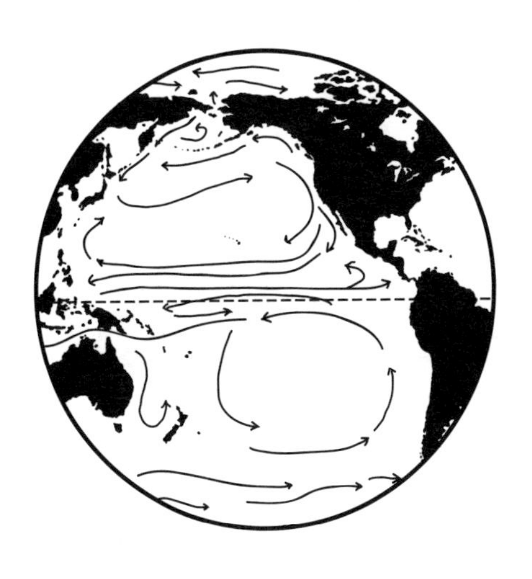

9. At this point, you may want to mark the currents in the Pacific on the overhead of the Pacific Rim Map. Alternatively, students can compare the currents made in the model ocean with those in the actual ocean as shown on the World Map of Currents sheet from Activity 1.

Introducing an Obstacle

Before beginning this activity, you may wish to ask students what caused the gyres in the previous activity. They should note two main factors: wind and the sides of the pan. The pan sides are obstacles themselves.

1. Move the overhead projector so it now projects onto the blank butcher paper or area of the chalkboard.

2. Ask students to think back on the waste disposal activity, and how the wind-driven currents were affected by continents that protected certain areas. Encourage them to use this prior experience to predict how they think the currents will be affected by an island (rock) sticking up in the path of the current.

3. Draw a rock outline on an area of the butcher paper or board where you will not be projecting and encourage them to show and perhaps draw their predictions on it.

4. Have the next student with a straw positioned at the surface as before, **but this time have the student on the side opposite the rock and blow towards the rock.** You will see turbulence (complicated and constantly changing swirling motion) as well as eddies and countercurrents forming around the island.

5. Point out how complicated currents can become in areas where there are many obstacles.

6. Working in their group of four, challenge your students to find gyres on their small inflatable globes. You may want to let them mark them with the overhead pens.

7. Remind them that the world is full of islands and irregular shapes like the rock just shown, unlike the smooth edge of the container used in the previous demonstration, and that other factors enter to complicate current patterns.

8. Ask a team to share one gyre they located. Ask which other teams found the same gyre.

9. Continue to do this with other gyres on the globe. Point out areas on the globe where obstacles have created strange currents, and where the gyres have been altered.

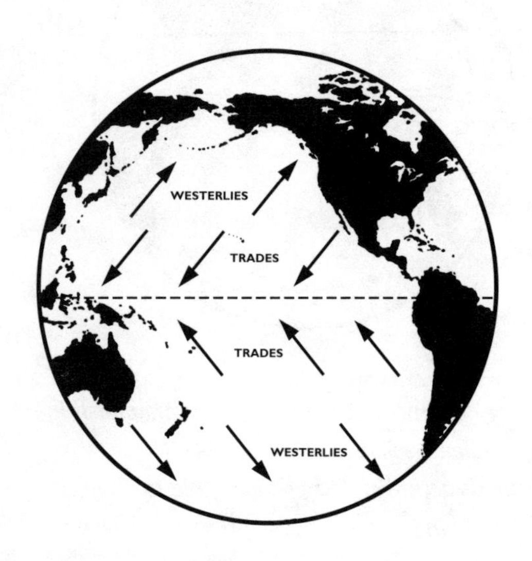

10. At this point you might lead a class discussion about the limitations and advantages of using models to simulate what happens in the real world.

11. In closing the activity, hold up the remaining two Key Concepts, and have one or more students read them aloud. Briefly discuss how these statements review the important ideas from today's activity. Post the Key Concepts on the wall near the first one and your posters for students to revisit during the rest of the unit.

- **Wind and the temperature differences between masses of water are two factors that cause currents.**
- **Winds blowing across the surface of the ocean—combined with other factors—cause major circulating currents, or gyres.**

Going Further

1. Research how wind-driven currents are mapped on the surface and from space. Investigate how deep ocean currents are mapped. The Internet is a wonderful resource for colored images of ocean temperatures and currents; the NOAA and NASA Web sites are especially useful. (See "Resources" on page 145.)

2. Oceanographers use the amount of dissolved oxygen in the water to determine the identity of currents and water masses. Use dissolved oxygen test kits such as those available from LaMotte to test different water samples. (See "Resources" on page 145.)

3. Have students work in pairs to write their own Key Concepts that represent things they learned and want to remember. Students can take charge of their own learning by deciding for themselves what they think is important. These new Key Concepts can also become part of student portfolios used for assessment.

Student Explorations with Waste Disposal

1. List at least four observations your team made about the waste (food coloring) traveling around in your ocean. One observation should describe how your results compared with the results of the controlled procedure. *Each team member should contribute at least one observation.*

2. What questions do you now have about what you observed? Some of your questions could start with the phrase "What would happen if . . . ?" *Each team member should contribute at least one question.*

3. Based on your observations, write at least two generalizations about how waste travels around the ocean.

Waste Disposal

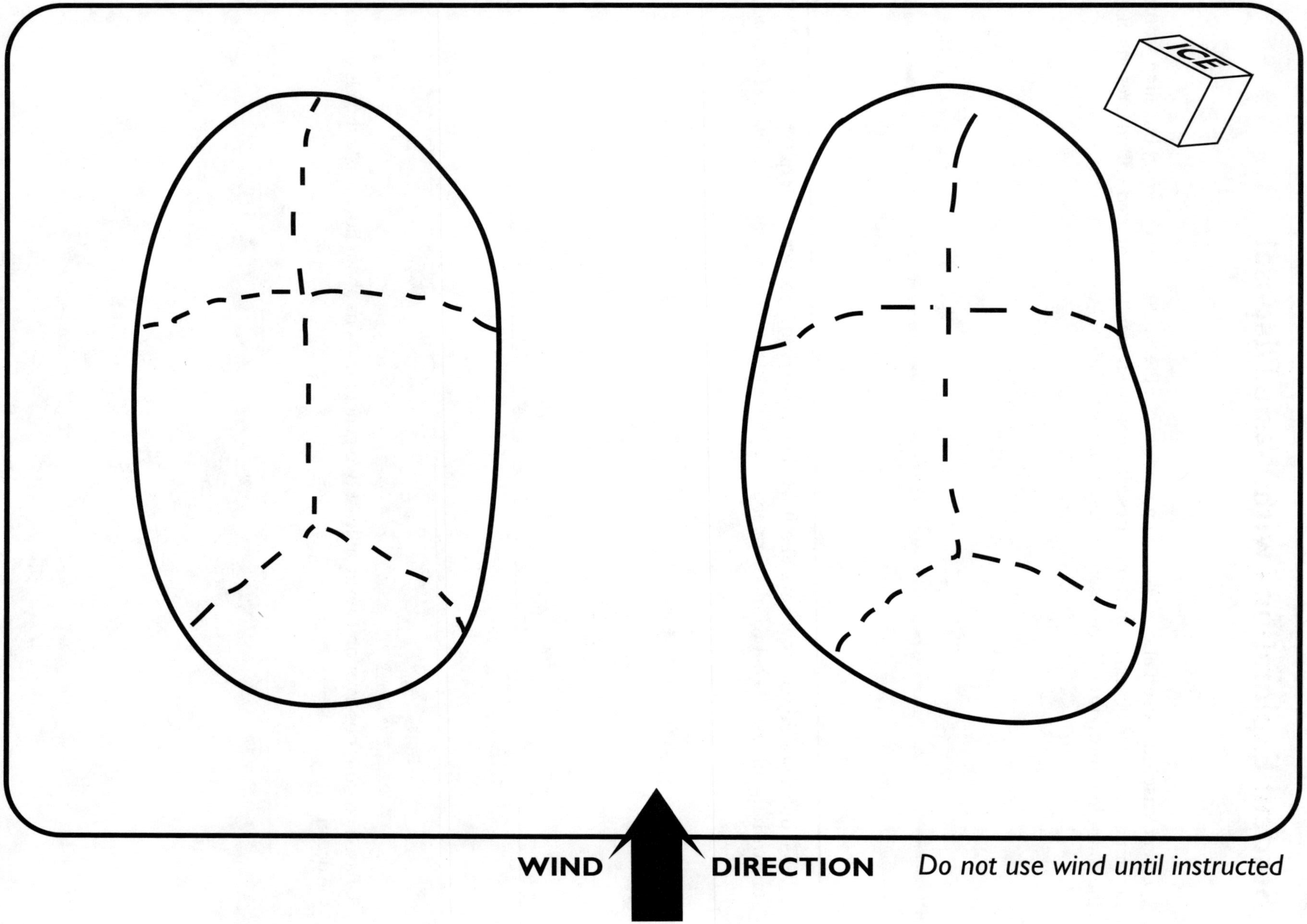

ICE

WIND DIRECTION

Do not use wind until instructed

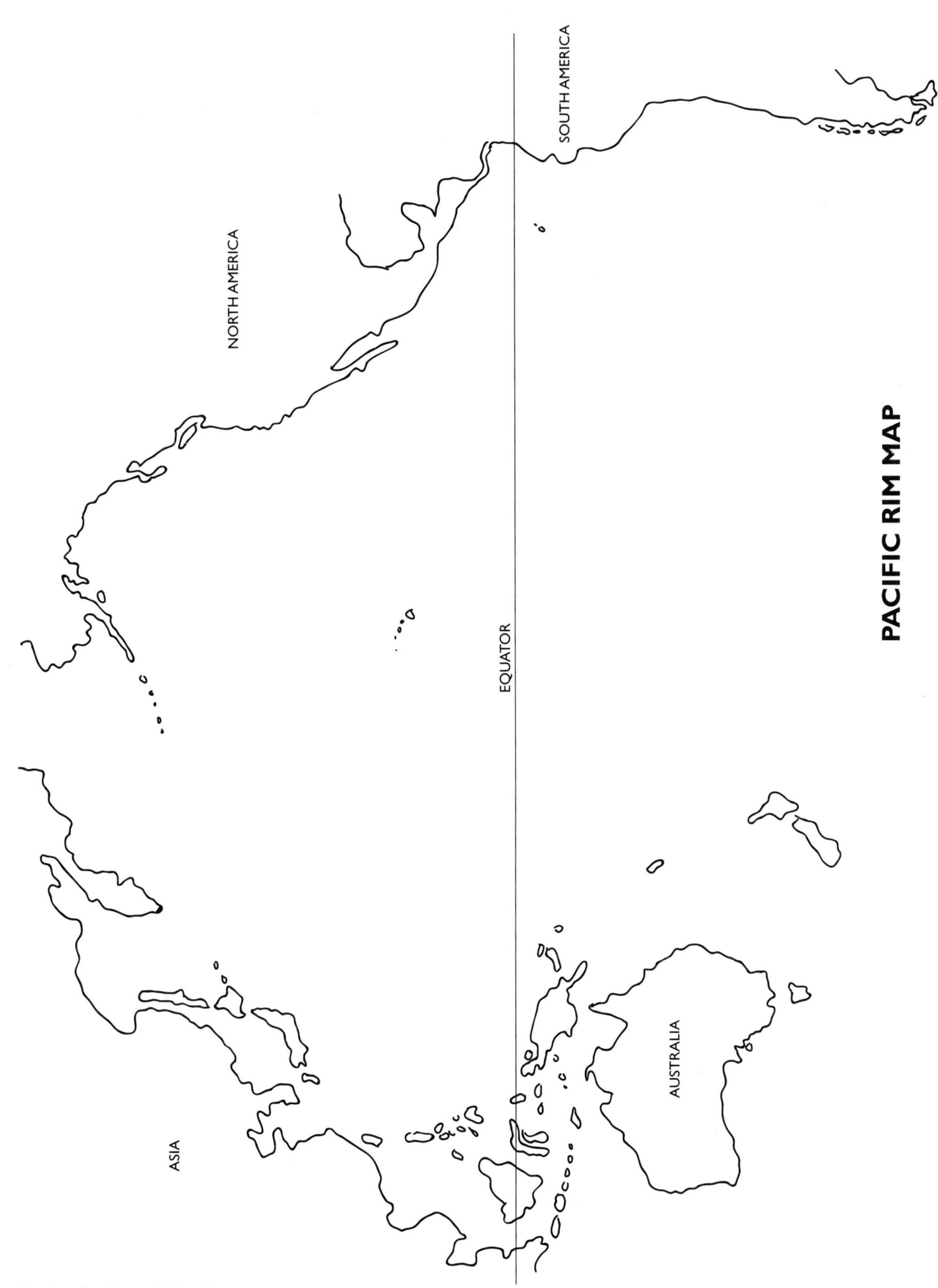
PACIFIC RIM MAP
NORTH AMERICA
SOUTH AMERICA
EQUATOR
ASIA
AUSTRALIA

2A

2C

Activity 3: Current Trends— Station Rotations

Overview

Students examine the relationship between temperature, salinity, and density. Cooperative student groups rotate through three different activities and experiments set up as stations. The students create currents by combining water of different temperature and salinity, and discover how the force of the wind and differences in density affect motion at all levels.

In Session 1, students complete two of the stations. In Session 2, they complete the third station and make a poster describing how it relates to actual currents. A discussion of these posters afterwards helps clarify the concepts and relates them to ocean currents.

The purposes of this activity:

- Provide students with a range of experiences relating to salinity and temperature
- Model how these factors and their interactions affect density and the creation of real currents

What You Need

For the class:

❒ 2–3 sheets of chart paper (approximately 27" x 34")
❒ 3–4 sets of markers (4–6 colors, wide tip)
❒ masking tape
Optional
❒ large, plastic, inflatable globe showing ocean currents or a wall map of the world

For each student:

❒ 1 Prediction data sheet, page 66
❒ pencil or pen

For each group of five students for Making Posters:

❒ 12" inflatable globe
❒ World Map of Currents from Activity 1, page 27
❒ sheet of chart paper
❒ markers (4–6 colors)

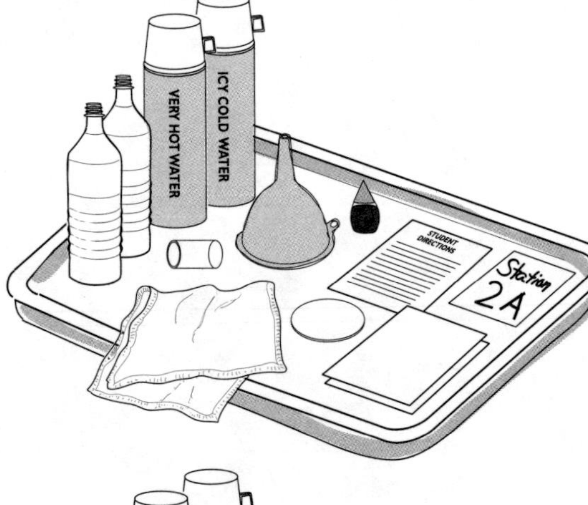

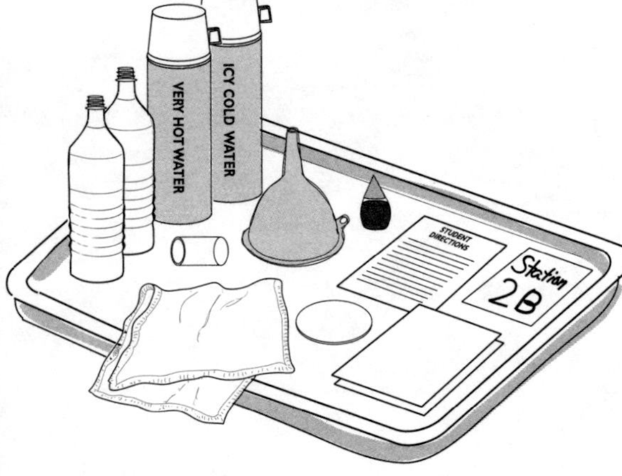

For each group of five students at the stations:

- ❒ 5 Current Trends data sheets, pages 67–69
- ❒ 5 pencils or pens
- ❒ markers (4–6 colors, include red and blue markers)

For the stations:

There are three different stations in the rotation. Each station is duplicated so two groups of students can work on the same station at the same time. The amounts listed below are for one station.

Stations 1A and 1B: Salinity Current Bottles

- ❒ station directions, page 63
- ❒ 2 identical bottles (12–16 oz. clear plastic, straight-sided water bottles with threaded mouths about 1" in diameter)
- ❒ room temperature tap water (enough to fill the bottles for each group)
- ❒ tornado tube (or Vortex Tube)
- ❒ dish towel (white preferred)
- ❒ ½ cup of salt (kosher preferred)
- ❒ vial (¼ fl. oz.) of dark food coloring (not yellow)
- ❒ tablespoon
- ❒ sheets of white paper
- ❒ tray
- ❒ yogurt lid with the rim cut off

Stations 2A and 2B: Temperature Current Bottles

- ❒ station directions, page 64
- ❒ 2 identical bottles (12–16 oz. clear plastic, straight-sided water bottles with threaded mouths about 1" in diameter)
- ❒ hot tap water to fill each group's bottle (about 100°F–110°F, which is not quite the hottest available from a home faucet)
- ❒ very icy-cold, refrigerated tap water to fill each group's bottle
- ❒ tornado tube (or Vortex Tube)
- ❒ dish towel (white preferred)
- ❒ vial (¼ fl. oz.) of dark food coloring (not yellow)
- ❒ sheets of white paper
- ❒ tray
- ❒ yogurt lid with the rim cut off

Optional

- ❒ funnel

Station 3A and 3B: Polar Versus Tropical Water

- ❒ station directions, page 65
- ❒ 2 (6–8 oz.) paper or Styrofoam cups
- ❒ 2 push pins
- ❒ container (approximately 6 quart, clear, rectangular, such as a plastic shoe box)
- ❒ 4 cups of hot (about 100°F–110°F) tap water
- ❒ 4 cups of very icy-cold, refrigerated tap water
- ❒ room temperature tap water (to fill large container about ¾ full)
- ❒ 20 marbles
- ❒ 2 (¼ fl. oz.) vials of food coloring (red and blue)
- ❒ spoon
- ❒ tray
- ❒ sheets of white paper

Getting Ready

1. Collect the plastic bottles with threaded mouths for Stations 1 and 2.

2. Kosher salt is preferred in this activity because when it dissolves it doesn't cloud the water like regular salt. About four tablespoons of salt is what you need for a 16 oz. bottle. **Note:** The appropriate amount of salt in the salinity currents station will create obvious currents almost immediately. Extra salt will not affect the outcome.

3. Determine how you will make and keep hot and cold water in your room. If your classroom doesn't have hot water, it works well to bring hot water in a thermos or boil tap water in an electric kettle or coffee maker. **Don't pour the boiling hot water directly into the plastic bottles or paper cups.** The boiling water will melt the plastic and burn hands. Mix the boiling water with just a little cold tap water so that it feels really hot, but doesn't burn. Better yet, place the boiling water in a thermos and allow to cool slightly before using or add just a little room temperature water to pre-mix it. If ice cubes are not available at your school, you can bring them in an ice chest from home to create very cold water.

4. Arrange the materials for each of the stations around the room. Place each of the station materials on a tray or in a separate container. Make a label for each station: Station 1A; Station 1B; Station 2A; Station 2B; Station 3A; Station 3B.

5. Make two copies of the Student Station Directions for each of the stations on colored paper; laminate if possible.

6. Duplicate the Prediction data sheet and Current Trends data sheets for each student; and the World Map of Currents for each group of five students.

7. Write the following for the Station Summary on a sheet of chart paper.

What is the name of the station?

Make a sketch of the station setup and your results.

Describe two observations your group made.

What is the Key Concept?

Where do you think the phenomenon shown by this station might occur in the real world?

Why?

What question(s) do you still have about this station?

If you were making up questions, what questions would you ask the class?

8. Write out the Key Concepts for this activity in large, bold letters on separate strips of chart paper and set aside.

- **Salinity and temperature differences create masses of water with different densities.**
- **Gravity causes more dense water to sink below less dense water. As a result, the less dense water rises.**

Session 1: Introducing the Stations and Making Predictions

1. Tell your students they will be rotating in small groups through a series of stations to help them gain a deeper understanding of ocean currents. Each group will complete two stations in Session 1, and a third station in Session 2.

2. Explain that for each station the A and B are identical, so it doesn't matter if they are assigned an A or a B for each of the three stations.

3. Briefly demonstrate or describe the experimental setup of Station 1 for the students. (Instructions for the stations are on the following pages.)

4. Distribute the Prediction data sheet to each student and have colored markers, pencils, or crayons available for the class. Describe what a prediction is, how to make predictions, and why it is important to do so. Briefly demonstrate how to use illustrations to record their predictions as described on their Prediction data sheet.

Tell the students that scientists make predictions as a way of explaining their ideas, and so others can test and improve their ideas.

5. Have the students make a prediction for Station 1—drawing on their Prediction data sheet where the colored water started and where they predict it will end up.

6. Demonstrate or describe the experimental setup for Stations 2 and 3—pausing after each of the descriptions to have students draw their predictions on their Prediction data sheet.

Demonstrating Station 1: Salinity Currents

Salinity refers to the amount of salt dissolved in water. In this experiment, students investigate the movement of colored water between bottles containing water with different amounts of salt.

1. Fill two bottles with room temperature tap water. Fill one bottle to the very top and leave about an inch of space at the top of the second bottle.

2. Add approximately four tablespoons of salt and six drops of food coloring to the bottle with space at the top and shake well. (This is the **salty** water bottle.)

3. Screw the tornado tube tightly onto the **salty** water bottle.

4. Finish filling the **salty** water bottle to the very top of the tornado tube with tap water.

5. Place the yogurt lid with the rim cut off over the top of the **fresh** water bottle.

6. Press down firmly on the yogurt lid, invert the **fresh** water bottle and quickly place it over the opening of the tornado tube attached to the **salty** water bottle. Carefully slide away the yogurt lid allowing the two bottles to join together with the tornado tube between them. Screw the fresh water bottle tightly into the tornado tube.

7. Lay the bottles gently on their side on a white dish towel to catch any drips. Try to disturb the bottles as little as possible—no jiggling, rolling, or shaking! Tighten the tornado tube if you see more than a few drops leaking from either bottle.

8. Bend down until your eyes are level with the bottles. Place a white piece of paper behind the bottles to make it easier to observe any movement of the colored water. Watch the movement for at least five minutes.

9. Answer the questions for Station 1 on a Current Trends data sheet.

10. Return the station to the way you found it for the next group.

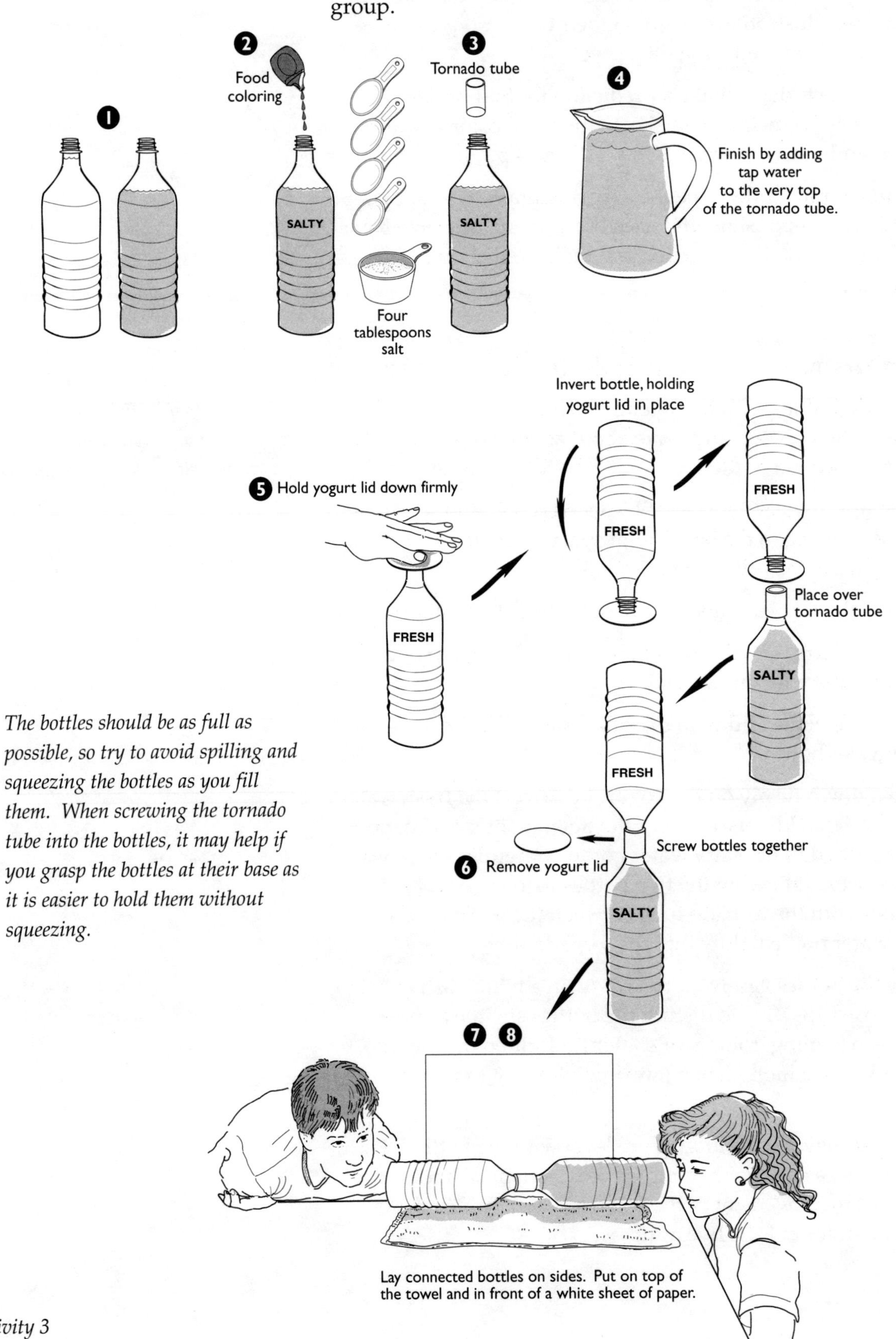

The bottles should be as full as possible, so try to avoid spilling and squeezing the bottles as you fill them. When screwing the tornado tube into the bottles, it may help if you grasp the bottles at their base as it is easier to hold them without squeezing.

Demonstrating Station 2: Temperature Currents

1. Fill one bottle to the very top with **hot** tap water.

2. Screw the tornado tube onto the **hot** water bottle.

3. Fill the other bottle almost to the top with icy **cold** tap water. (The space at the top is so food coloring can be added and mixed in.)

4. Add six drops of food coloring to the **cold** bottle, and shake well.

5. Finish filling the **cold** water bottle to the very top with more cold water.

6. Place the yogurt lid with the rim cut off over the top of the **cold** water bottle so there is less chance for spillage when you turn it upside down to place it on top of the **hot** water bottle.

7. Press down firmly on the yogurt lid, invert the **cold** water bottle and quickly place it over the opening of the tornado tube attached to the **hot water** bottle. Carefully slide away the yogurt lid allowing the two bottles to join together with the tornado tube between them. Screw the **cold** water bottle tightly into the tornado tube.

8. Lay the bottles gently on their side on a white dish towel to catch any drips. Try to disturb the bottles as little as possible—no jiggling, rolling, or shaking! Tighten the tornado tube if you see more than a few drops leaking from either bottle.

9. Bend down until your eyes are level with the bottles. Place a white piece of paper behind the bottles to make it easier to observe any movement of the colored water. Watch the movement for at least five minutes. **Be sure to also feel the bottles for differences in temperature. Also, look for other signs of temperature difference, such as condensation.**

10. Answer the questions for Station 2 on your Current Trends data sheet.

11. Return the station to the way you found it for the next group.

In this experiment, students investigate the movement of colored water between bottles containing different temperatures of water.

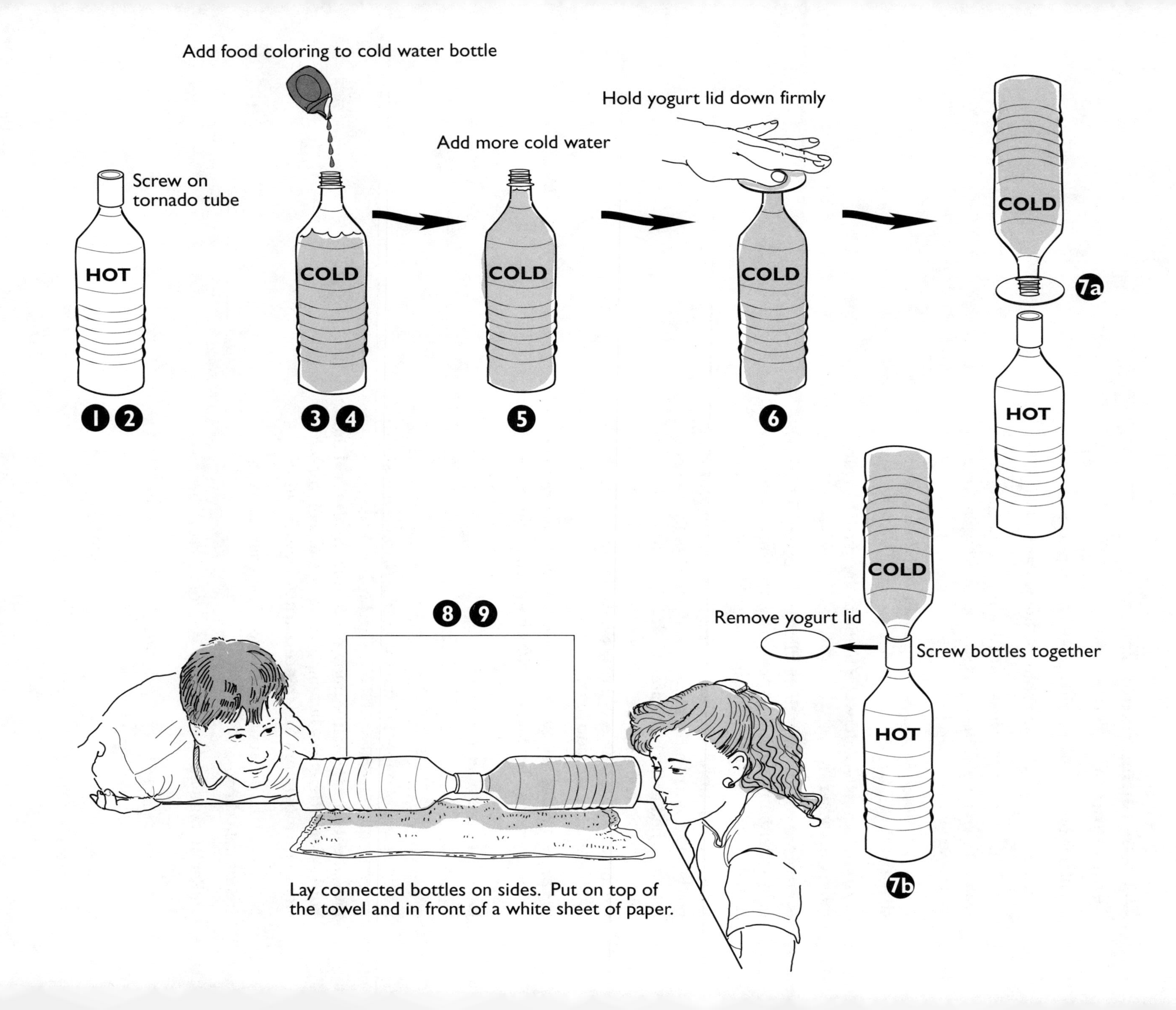

Screw on tornado tube
HOT
1 2
Add food coloring to cold water bottle
COLD
3 4
Add more cold water
COLD
5
Hold yogurt lid down firmly
COLD
6
COLD
HOT
7a
COLD
Remove yogurt lid
Screw bottles together
HOT
7b
8 9
Lay connected bottles on sides. Put on top of the towel and in front of a white sheet of paper.

Demonstrating Station 3: Polar Versus Tropical Water

In this activity, students compare the movement and position of three different temperatures of water.

1. Place marbles in the cups to keep them from floating or tipping.

2. Determine the exact level of water to add to the container by comparing it to the height of the cup you are using. Place one of the cups in the container for a reference and add room temperature water to the container to a height about half an inch below the rim of the cup.

3. Remove the cup from the container. Pour icy-cold water in one of the cups and add six drops of blue food coloring. Stir carefully with the spoon.

Linking color with cold, salty, and hot reinforces that it is those properties, not color, that determines density and flow.

4. Pour very hot water in the other cup and add six drops of red food coloring. Stir carefully with the spoon.

5. Stick a push pin in each cup at a level where the hole is just below the surface of the water in the large container. The push pins should be at the same level in both cups. Leave the push pins in the cups!

6. Carefully place the cups in the container of water with the push pins facing away from each other. Your setup should look something like the diagram below.

You may want to include some sort of bonus points, or other reward for the groups where you see particular social skills being used. Some appropriate and practical ones for this session include checking for understanding, cooperation, and encouraging. At first they may "fake it" when they see you nearby, but with practice it will become routine and powerful as they realize that each of them has something to offer to each other.

7. Pull out the push pin in each cup. Bend down so your eyes are level with the setup. Place a sheet of white paper behind the container so any water movement is easily seen.

8. Continue to add water to the two cups (hot water to the hot cup and cold water to the cold cup) to keep the water level in each cup almost to the top.

9. Answer the questions for Station 3 on your Current Trends data sheet.

10. Return the station to the way you found it for the next group.

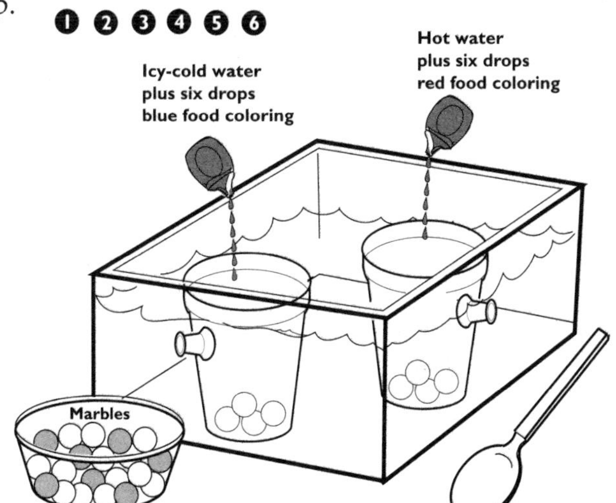

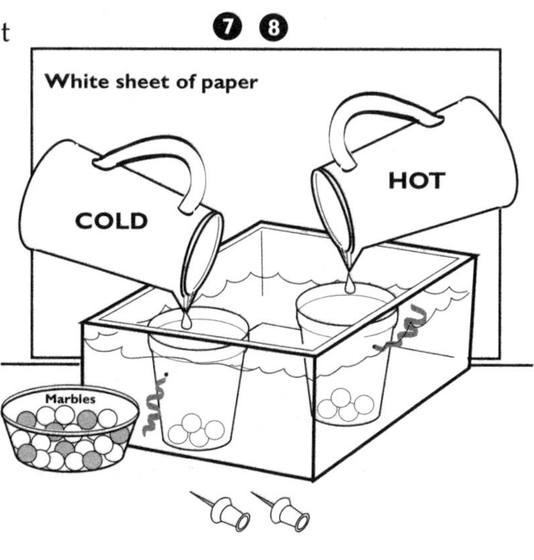

Station Rotations

1. Assign each student to one of six groups. Assign each group to a different station.

2. Pass out the Current Trends data sheets, one per student. Tell them the first step is to have one of them read the station directions aloud.

3. Ask the students to divide between themselves the roles needed to perform their tests. Stress that everyone needs to participate in the activity and fill in their own data sheet—although consultation with others is encouraged.

4. Tell the students when you say to rotate they should go to the next higher numbered station. Those at Station 3 will go to Station 1. Give students 15 minutes to complete each of the two stations. If they complete a station quicker, suggest they finish any incomplete questions on their data sheet from a previous station.

5. Proceed with the station rotations. After each group has been to two of the stations, provide time for group members to compare their results and complete their data sheets. Some groups may want to repeat a test if their initial observations were not detailed enough to answer the questions on the data sheet.

Session 2: Completing the Stations

1. Have the students complete the station rotations by doing their third (and final) station. Provide time for group members to compare their results and complete their data sheets.

2. Tell the groups they will be asked to prepare a poster about this station to present to the rest of the class.

Making Station Posters

1. Tell the students they will discuss and review the results of the stations and relate their observations to the real-world ocean current system. They will work with their station rotation group to prepare a poster to present the results of their final station. (Have them keep their last station materials and setup to use as props for their presentation.)

2. Distribute the World Map of Currents and the globes to each of the six station groups. Also distribute the chart paper and markers to make the posters. Have available the large globe or wall map. During the following activities, have students locate the places described.

3. Tape the Station Summary chart on the board and have each of the groups use the chart as a guide to prepare a poster. Have one person in each group act as the recorder and write down their responses on the poster.

What is the name of the station?

Make a sketch of the station setup and your results.

Describe two observations your group made.

What is the Key Concept?

Where do you think the phenomenon shown by this station might occur in the real world?

Why?

What question(s) do you still have about this station?

If you were making up questions, what questions would you ask the class?

4. When each group is finished, have them tape the posters around the room and give a short summary of their poster. The rest of the class can ask them any questions they recorded on their own Current Trends data sheet about this station.

You may want to lead a brief discussion about salinity, depending on how much prior knowledge students demonstrate. Salinity can be described as the "saltiness" of the water, and these experiments use table salt (NaCl or sodium chloride). However, when seawater evaporates many other "salts" besides NaCl are present, including magnesium chloride, magnesium sulfate, potassium chloride, and calcium carbonate. NaCl is the major ingredient and makes up slightly more than 78% of the total salt. Salt water aquarists use something called "Instant Ocean" to make sea water for their aquariums if they can't get it from the actual sea. This product replicates ocean water and uses all the different constituents of sea water.

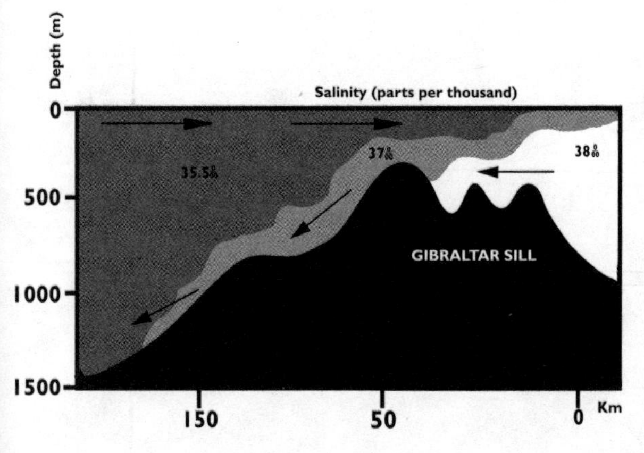

5. Leave the posters up around the room and tell the students they might want to refer to them during the following class discussion and wrap-up.

Debriefing the Station Posters

Salinity Currents and Layering

1. Read the observations and the Key Concept from the student posters.

2. Ask the students what they think causes the water to move in salinity currents. [Because of gravity, water that is more dense (saltier) tends to sink and that causes water that is less dense (less salty) to rise. This water movement causes salinity-related density currents.]

3. Ask where they think salinity currents might occur and why.

4. If they didn't mention the Mediterranean Sea, have them find it on a map or globe. The Mediterranean is very salty because the warm weather causes a lot of evaporation, leaving behind very salty warm water. There are also relatively few fresh water rivers emptying into it.

5. Ask the students where they think this salty water might go. Does it mix with the Atlantic, float on the surface, or sink below? [It sinks down and flows out into the Atlantic Ocean at the Strait of Gibraltar. Less salty water entering the Mediterranean Sea from the Atlantic floats on top of this denser layer of water and forms the surface water layer in the western Mediterranean.]

6. Point out that although the water is salty, which would make it dense, it is also warm, which makes it less dense. Ask students where they think it would end up in the layers of the Atlantic: bottom, middle, or top. [It has a medium density and tends to remain at a middle depth.]

7. Have the students find the Amazon River, and where it flows into the ocean. Point out that the outflow from the Amazon is huge, and ask what they think might happen to all that fresh water when it reaches the ocean. [The fresh river water floats on top of the ocean water. Early European explorers of South America were amazed when they found a surface layer of fresh water far out at sea. They followed the fresh surface layer to its source, and "discovered" the Amazon River.]

Temperature Currents and Layering

1. Read the observations and the Key Concept from the student posters.

2. Ask students what they think causes the water to move in temperature currents. [Because of gravity, water that is more dense (cold) tends to sink, which causes water that is less dense (warm) to rise. This water movement causes temperature-related density currents.]

3. Ask if they have ever noticed this phenomenon themselves while swimming in a lake or pool. The surface waters are warmed by the sun and the depths remain cold. Ask what time of year would it probably be most evident. [Summer, when the sun heats up the surface waters.]

4. Ask why the warm water remains on top, and the cold water sinks below. [Warm water is less dense than cold water.] Layers separated by temperature can stay that way for a long time, because the more dense, colder water stays cold, and the less dense, warmer water at the surface stays warmer. Students might wonder why the warm water doesn't cool down and the cold water heat up as they are next to each other. Actually, that would eventually happen, but given the huge mass of water in the ocean it would take years. Even in the bottle experiments it may take an hour or so for the temperatures of the water to equalize.

5. Ask what time of year cold water might be found on top, and warmer water below. [In the winter, when the air temperature becomes colder than the water temperature.] Will the water temperature stay layered this way? What will happen? Ask where they think temperature currents might occur in the real world. [The cold water will sink and the warm water will rise forming a temperature current. This happens a lot in the fall and winter in mid- and higher latitudes. It can lead to "overturning" in lakes, often detected by smell as the lower warm bottom waters rise to the surface bringing with them hydrogen sulfide and other smelly decomposition gases.]

The zone in which temperature decreases rapidly with depth is called the thermocline. Below the thermocline the temperature is usually pretty cold and changes relatively little with depth.

6. Explain that with salinity and temperature differences combined, different water masses don't mix easily when they meet, but instead flow above or beneath each other. These differences can last a very long time—up to 1,600 years in some cases!

7. Ask your students how these layers might affect the animals living in the ocean. [This layering means that there may be two—or more—different habitats for organisms to live in just in one column of water. The lower water may be salty, cold, and dense. The upper layer of water may be less salty, warm, and less dense.]

8. Explain that the upper layers of the ocean are where plantlike organisms (phytoplankton and other algae) live, because they need sunlight, but the "fertilizer" or nutrients from dead, decomposed organisms needed for growth sink toward the bottom and collect in the deeper, colder, and denser water. If these nutrient-rich waters are brought back to the surface through upward currents, they can create areas rich with life.

Polar Versus Tropical Water

1. Read the observations and the Key Concept from the student posters.

2. Ask where they think the phenomenon shown by this station might occur in the real world. Why?

3. Although layers in the ocean may sometimes last a long time, there is also a lot of mixing between the layers. Ask what students think may cause the layers to mix. [Answers will vary, but may include—if surface water becomes more dense, perhaps by cooling, it will sink, and the less dense water will rise to the surface. These density currents could then cause mixing between the layers. Also, there may be mixing by waves especially during large storms.]

4. Ask if anyone noticed dense, cold water move along the bottom then hit the side of the container and flow back up to the surface. This phenomenon in the ocean is called ***up-welling.*** Ask why this might be important to organisms living in the ocean. [Upwelling brings cold nutrient-rich water to the surface to act as fertilizer for the phytoplankton living in the sunlit surface waters (photic zone).] Let them know upwelling zones are usually very rich with life.

5. Explain that winds are usually the cause of coastal up-welling by moving the surface waters offshore. These wind currents are acted upon by other forces which move the less dense, warm surface water away, leaving a deficit of water along the shore and allowing colder, more dense water from below to come up (well up) to the surface.

6. Tell your students most wind-driven currents occur in about the top 20% of the ocean. The movement in the other 80% of the ocean is caused mainly by density differences. The major cause of density currents throughout the ocean is the warming and cooling of ocean water at the surface, and the changes in the salinity of the surface water.

7. Hold up the Key Concepts for this activity, one at a time and have one or more students read them aloud. Post the concepts on the wall for students to revisit during the rest of the unit.

- **Salinity and temperature differences create masses of water with different densities.**
- **Gravity causes more dense water to sink below less dense water. As a result, the less dense water rises.**

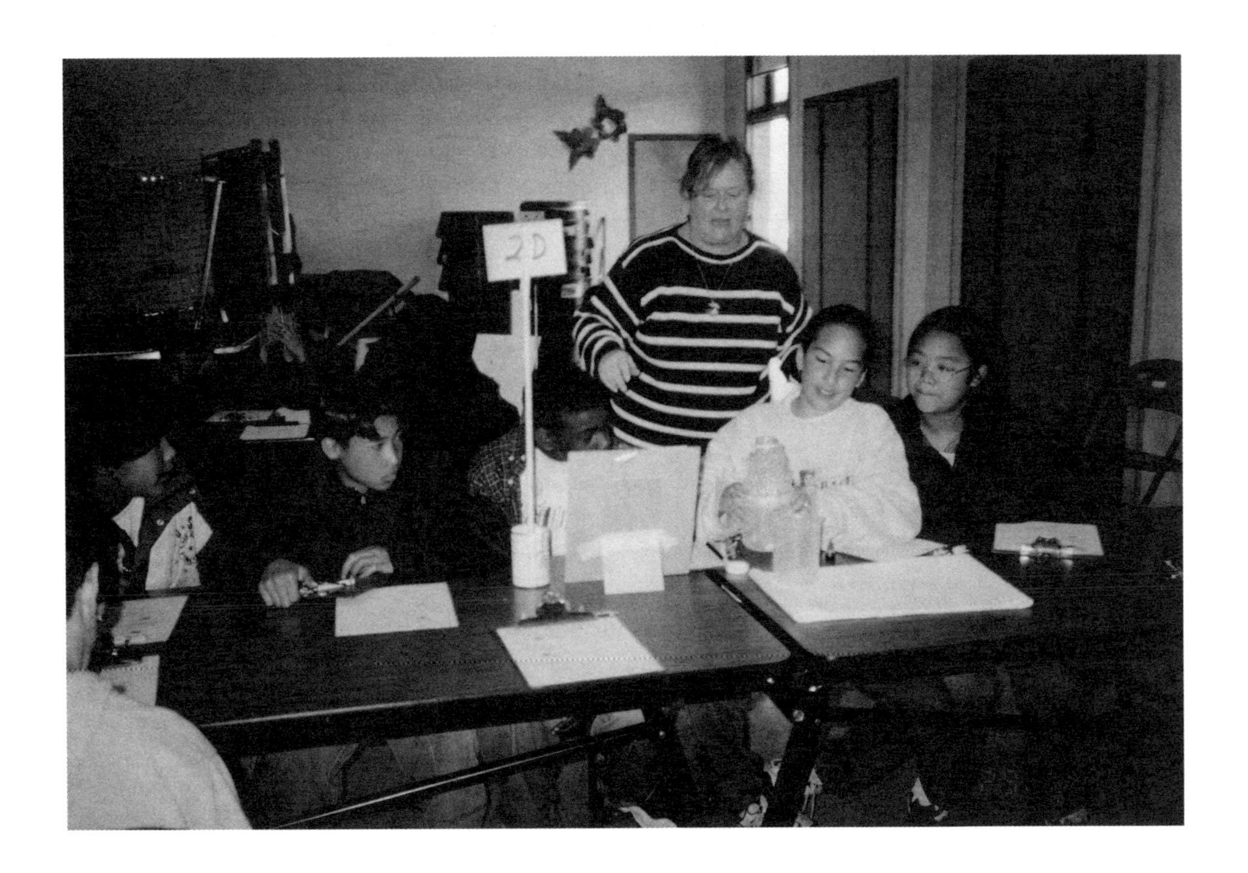

Going Further

1. Hydrometers

Make your own hydrometer to test the relative density of each of the different water samples in this activity. This gives a more quantitative measure of density than the colored water layers and currents.

How to Make a Hydrometer

Use a plastic straw, a waterproof thin-point marking pen, and a small ball of plasticine clay. Make the hydrometer by sealing one end of the straw with a small amount of clay and then drawing lines on the straw every 2 mm. Float the hydrometer in fresh water and note the level at which it floats; do the same in salt water and compare. You can also make a hydrometer with a new unsharpened pencil with a tack stuck into the eraser. Mark on the pencil every 2 mm with a waterproof marking pen.

2. Putting Your Knowledge to Work: The Unknown

Provide students with a water sample of unknown salinity and have students determine whether it is fresh water, slightly salty, or very salty by comparing it with other samples distributed to the group. Students can use any of the methods they tried at the stations. **Remember to remind students they should never taste unknown solutions.** They can then fill out a form, such as:

1. My group #__________ My sample #_________

2. These are the steps I took to determine whether I have fresh, slightly salty, or very salty water:

3. My sample is (circle one)

fresh water slightly salty water very salty water

4. Here is the evidence for my conclusion:

Station Directions

Station 1: Salinity Currents

Salinity refers to the amount of salt dissolved in water. In this experiment, you investigate the movement of colored water between bottles containing water with different amounts of salt.

1. Fill two bottles with room temperature tap water. Fill one bottle to the very top and leave about an inch of space at the top of the second bottle. Try to avoid spilling and squeezing the bottles as you fill them.

2. Add approximately four tablespoons of salt and six drops of food coloring to the bottle with space at the top and shake well. (This is the **salty** water bottle.)

3. Screw the tornado tube tightly onto the **salty** water bottle.

4. Finish filling the **salty** water bottle to the very top of the tornado tube with tap water.

5. Place the yogurt lid with the rim cut off over the top of the **fresh** water bottle.

6. Press down firmly on the yogurt lid, invert the **fresh** water bottle and quickly place it over the opening of the tornado tube attached to the **salty** water bottle. Carefully slide away the yogurt lid allowing the two bottles to join together with the tornado tube between them. Screw the fresh water bottle tightly into the tornado tube. When screwing the tornado tube into the bottles, it may help if you grasp the bottles at their base as it is easier to hold them without squeezing.

7. Lay the bottles gently on their side on a white dish towel to catch any drips. Try to disturb the bottles as little as possible—no jiggling, rolling, or shaking! Tighten the tornado tube if you see more than a few drops leaking from either bottle.

8. Bend down until your eyes are level with the bottles. Place a white piece of paper behind the bottles to make it easier to observe any movement of the colored water. Watch the movement for at least five minutes.

9. Answer the questions for Station 1 on a Current Trends data sheet.

10. Clean up and prepare the station for the next group.

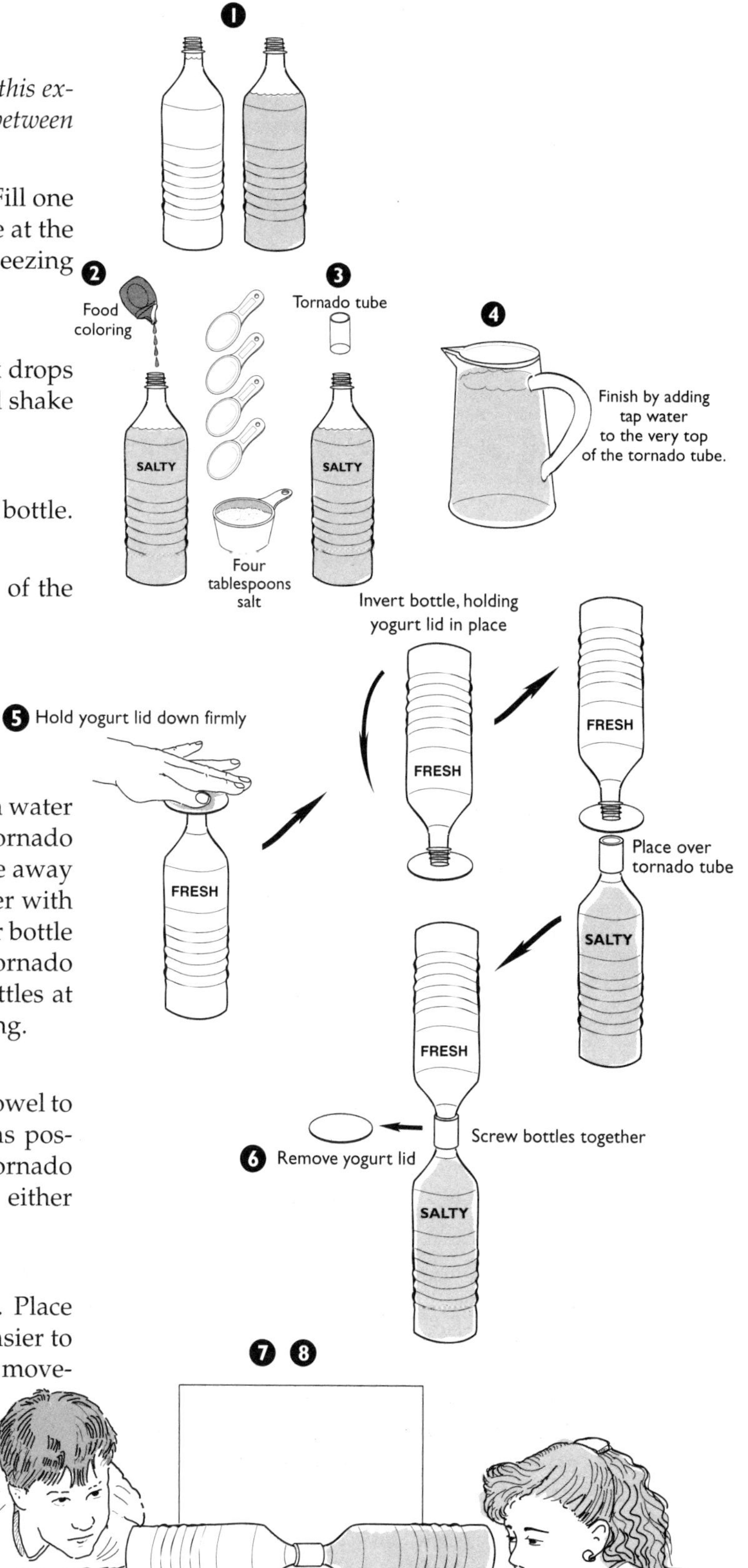

Lay connected bottles on sides. Put on top of the towel and in front of a white sheet of paper.

Station Directions

Station 2: Temperature Currents

In this experiment, you investigate the movement of colored water between bottles containing different temperatures of water.

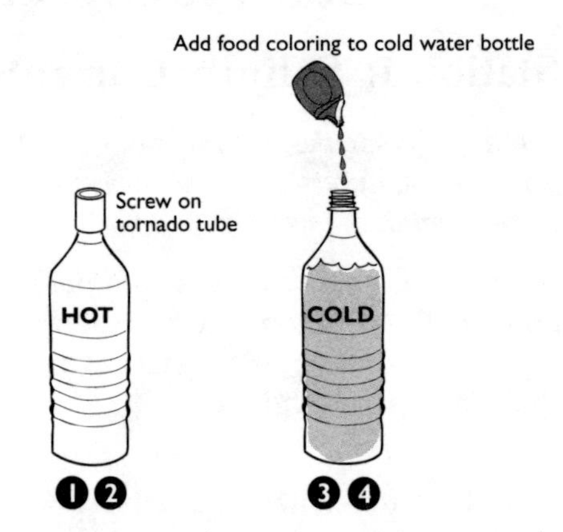

1. Fill one bottle to the very top with **hot** tap water.

2. Screw the tornado tube onto the **hot** water bottle.

3. Fill the other bottle almost to the top with icy **cold** tap water. (The space at the top is so food coloring can be added and mixed in.) Try to avoid spilling and squeezing the bottles as you fill them.

4. Add six drops of food coloring to the **cold** bottle, and shake well.

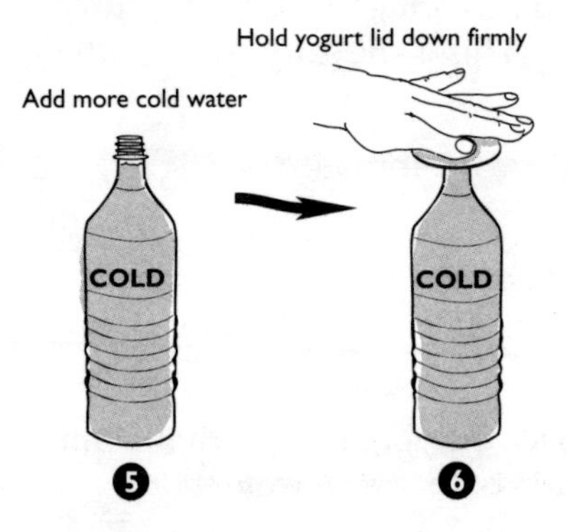

5. Finish filling the **cold** water bottle to the very top with more cold water.

6. Place the yogurt lid with the rim cut off over the top of the **cold** water bottle so there is less chance for spillage when you turn it upside down to place it on top of the **hot** water bottle.

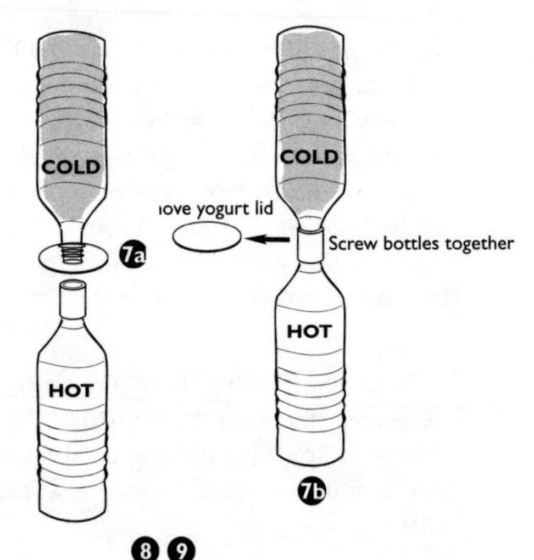

7. Press down firmly on the yogurt lid, invert the **cold** water bottle and quickly place it over the opening of the tornado tube attached to the **hot water** bottle. Carefully slide away the yogurt lid allowing the two bottles to join together with the tornado tube between them. Screw the **cold** water bottle tightly into the tornado tube. When screwing the tornado tube into the bottles, it may help if you grasp the bottles at their base to hold them without squeezing.

8. Lay the bottles gently on their side on a white dish towel to catch any drips. Try to disturb the bottles as little as possible—no jiggling, rolling, or shaking! Tighten the tornado tube if you see more than a few drops leaking from either bottle.

Lay connected bottles on sides. Put on top of the towel and in front of a white sheet of paper.

9. Bend down until your eyes are level with the bottles. Place a white piece of paper behind the bottles to make it easier to observe any movement of the colored water. Watch the movement for at least five minutes. **Be sure to also feel the bottles for differences in temperature. Also, look for other signs of temperature difference, such as condensation.**

10. Answer the questions for Station 2 on your Current Trends data sheet.

11. Return the station to the way you found it for the next group.

Station Directions

Station 3: Polar Versus Tropical Water

In this activity, you compare the movement and position of three different temperatures of water.

1. Place marbles in the cups to keep them from floating or tipping.

2. Determine the exact level of water to add to the container by comparing it to the height of the cup you are using. Place one of the cups in the container for a reference and add room temperature water to the container to a height about half an inch below the rim of the cup.

3. Remove the cup from the container. Pour icy-cold water in one of the cups and add six drops of blue food coloring. Stir carefully with the spoon.

4. Pour very hot water in the other cup and add six drops of red food coloring. Stir carefully with the spoon.

5. Stick a push pin in each cup at a level where the hole is just below the surface of the water in the large container. The push pins should be at the same level in both cups. Leave the push pins in the cups!

6. Carefully place the cups in the container of water with the push pins facing away from each other. Your setup should look something like the diagram below.

7. Pull out the push pin in each cup. Bend down so your eyes are level with the setup. Place a sheet of white paper behind the container so any water movement is easily seen.

8. Continue to add water to the two cups (hot water to the hot cup and cold water to the cold cup) to keep the water level in each cup almost to the top.

9. Answer the questions for Station 3 on your Current Trends data sheet.

10. Return the station to the way you found it for the next group.

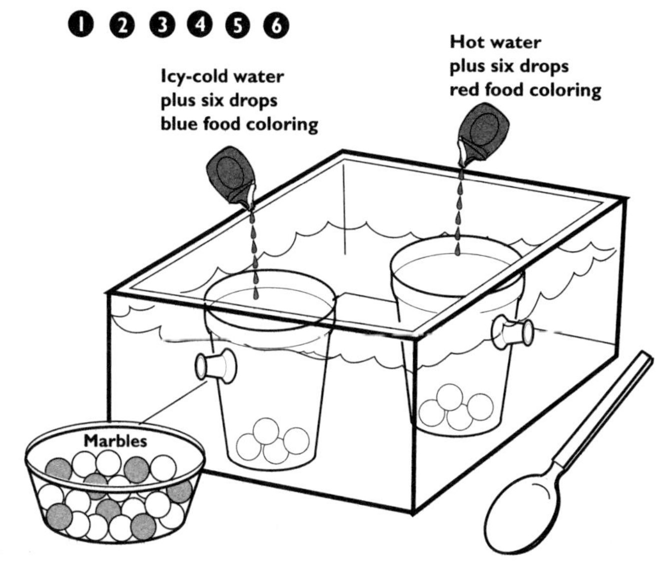

Student Prediction Data Sheet

Station 1: Salinity Currents

Use colored markers on the illustrations below to mark your prediction about the movement of the colored, salty water in the bottles, and where it will end up. Label which bottle has the fresh water and which bottle has the salty water at the start of the exploration.

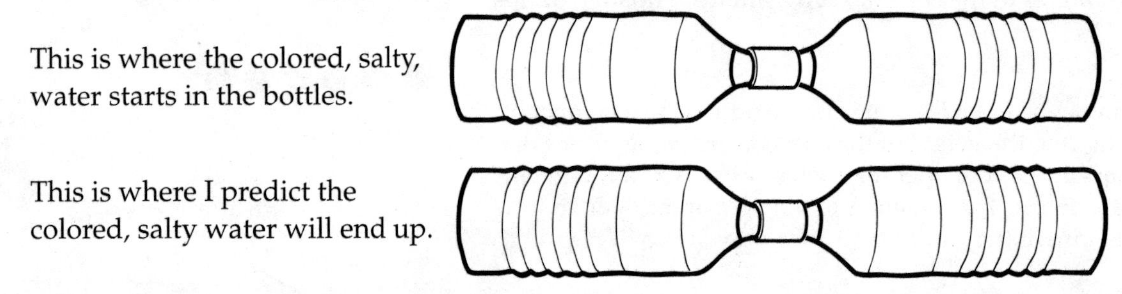

This is where the colored, salty, water starts in the bottles.

This is where I predict the colored, salty water will end up.

Station 2: Temperature Currents

Use colored markers on the illustrations below to mark your prediction about the movement of the colored, cold water in the bottles and where it will end up. Label which bottle has the cold water and which has the hot water at the start of the exploration.

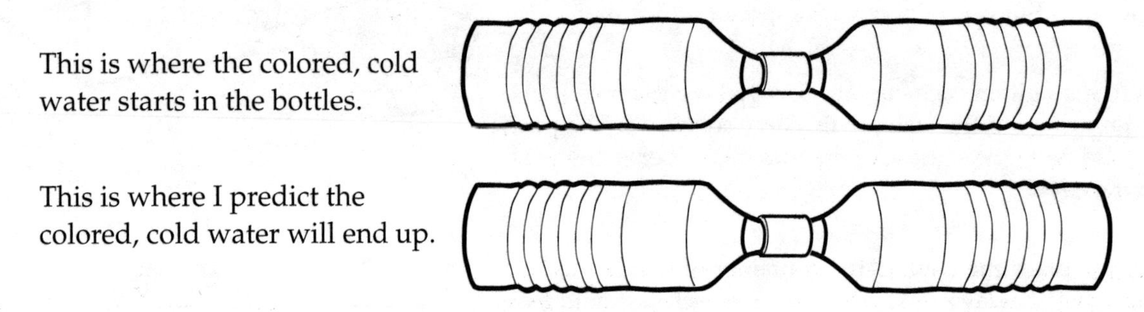

This is where the colored, cold water starts in the bottles.

This is where I predict the colored, cold water will end up.

Station 3: Polar Versus Tropical Water

Using colored markers and the diagrams below, mark your prediction about where you think the colored waters will end up as the room temperature water, cold water, and hot water meet. Label the diagrams and use a different color for each of the different temperatures of water.

This is where the blue (cold) water and red (hot) water was at the start.

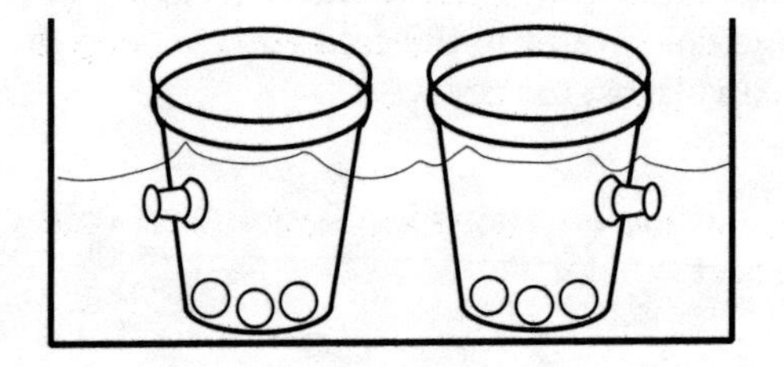

This is where I predict the blue (cold) water and red (hot) water will end up.

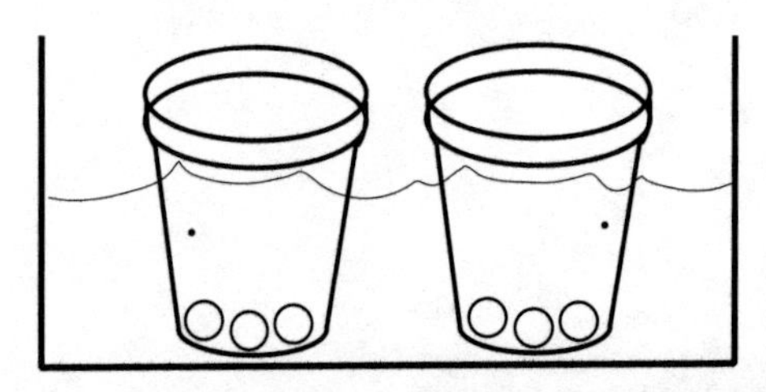

Station 1: Salinity Currents

1. Use colored markers on the illustrations below to show the movement of the colored water in the bottles and where it ended up. Label the bottles on the diagram to show which has fresh water and which has salty water at the start of the exploration.

This is where the colored, salty water started in the bottles.

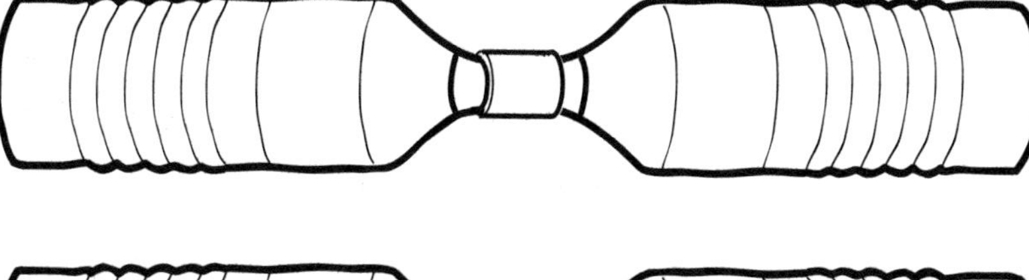

This is where the colored, salty water ended up in the bottles.

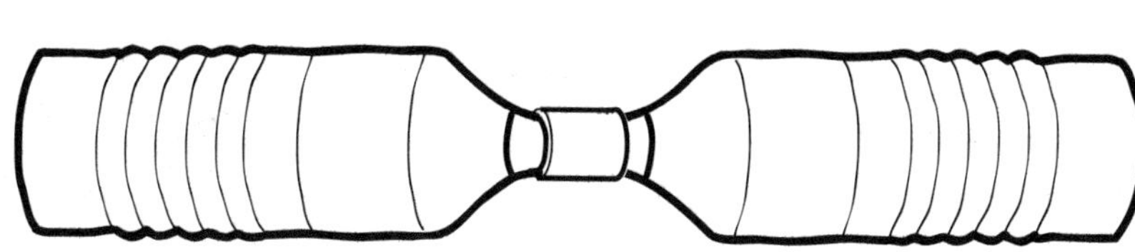

2. Briefly describe your results.
Include at least two things you observed.

3. How did your prediction compare with your results?
Were you surprised by the results? Why or why not? Be specific.

4. Based on your observations, describe at least one generalization your group could make about what happens when waters of different salinities meet.

5. Make up at least one question about the results of this station to ask the class. (You don't have to know the answer.)

Student Current Trends Data Sheet

Station 2: Temperature Currents

1. Use colored markers on the illustrations below to show the movement of the colored water in the bottles and where it ended up. Label the bottles on the diagram to show which has cold water and which has hot water at the start of the exploration.

This is where the colored, cold water started in the bottles.

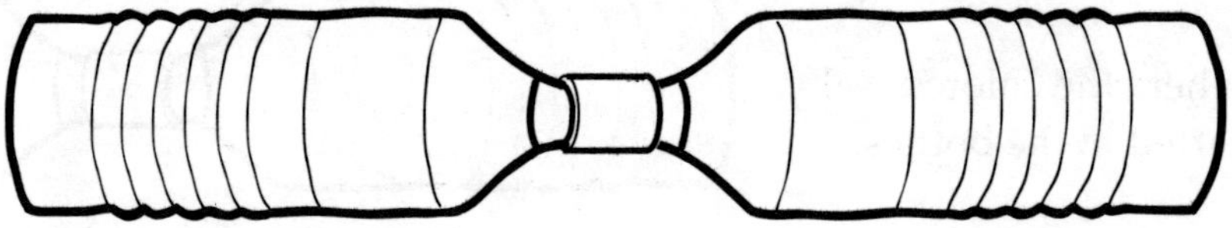

This is where the colored, cold water ended up in the bottles.

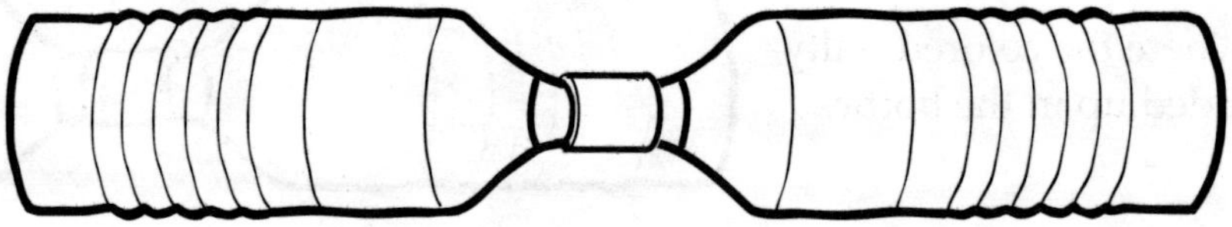

2. Briefly describe your results. Include at least two things you observed—including what you noticed when you felt the bottles for temperature differences.

3. How did your prediction compare with your results? Were you surprised by the results? Why or why not? Be specific.

4. Based on your observations, describe at least one generalization your group could make about what happens when water of different temperatures meet.

5. Make up at least one question about the results of this station to ask the class. (You don't have to know the answer.)

Station 3: Polar Verses Tropical Water

1. Using colored markers and the diagrams below, illustrate what happened to the different colors of water as you pulled the pins out of the cups. Show where the room temperature water, cold water, and hot water started and ended up. Label and use a different color for each of the different temperatures of water.

This is where the clear (room temperature) water, blue (cold) water, and red (hot) water started.

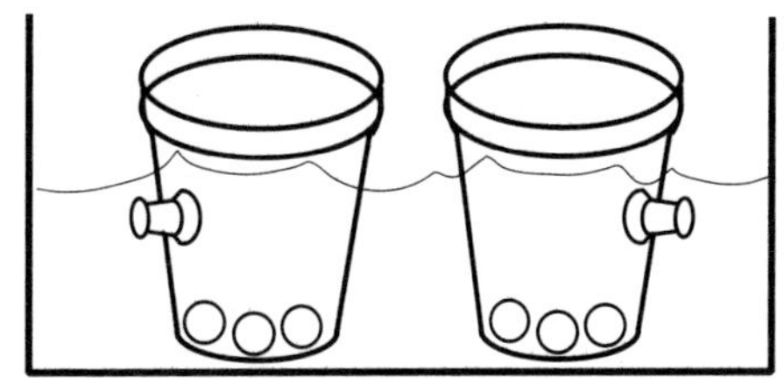

This is where the clear (room temperature) water, blue (cold) water, and red (hot) water ended up.

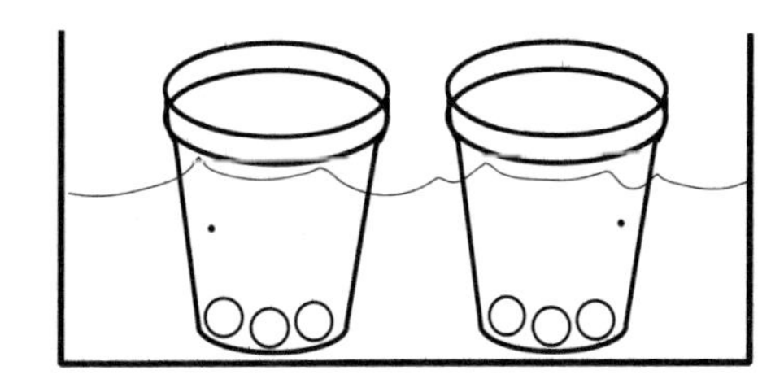

2. Briefly describe your results in words. Include at least two things you observed. One observation could be a description of what happens when the blue (cold) layer and the red (hot) layer reach the opposite side of the container.

3. How did your prediction compare with your results? Were you surprised by the results? Why or why not? Be specific.

4. Based on your observations, describe at least one generalization your group could make about what happens when water of different temperatures meet.
***Challenge Question**: Make a prediction about what sort of temperature and salinity layering you would expect to find if you swim after a summer rain when the ocean is calm.*

5. Make up at least one question about the results of this station to ask the class. (You don't have to know the answer.)

Activity 4: Layering Liquids

Overview

Students are challenged to apply information they have learned about different liquids to create four distinct layers in straw cylinders using only colored water and salt. In a follow-up discussion, the concept of density is introduced at a molecular level, and students are guided to an understanding that explains the concrete phenomena they have witnessed.

The purposes of this activity:

- Delve more deeply into the concept of density
- Introduce aspects of density at the molecular level
- Relate these new understandings to ocean currents

"I had so much fun watching the students' eyes light up and their minds roll as they tried their predictions. Kids really loved this one—we all enjoyed the variables and the problem-solving aspect."

"This activity really helped to deepen my students' understanding of the concept of density. It was a good assessment activity too—summarized the previous labs and even more importantly allowed the students to actually test their ***own*** *knowledge. They were really able to put to use what they had learned before—very empowering."*

What You Need

For the class:

- ❒ 8–10 cups of very hot tap water or the means to make it
- ❒ 4 cups of room temperature or cold water
- ❒ 4 thermoses with at least 3-cup capacity each
- ❒ sharp knife for slicing potatoes
- ❒ at least 8 ice cubes
- ❒ 4 (¼ fl. oz.) vials of food coloring (red, blue, yellow, and green)
- ❒ 3 cups of salt (kosher preferred)
- ❒ 1- or 2-cup clear measuring cup
- ❒ tablespoon
- ❒ 2 stir sticks
- ❒ a source of water
- ❒ 1 tablespoon of oil
- ❒ masking tape
- ❒ 3 sheets of chart paper (approximately 27" x 34")

For each group of 4–6 students for equipment tray:

- ❒ 2–3 clear plastic straws
- ❒ medium-sized raw potato, cut into 4–6 1" slices
- ❒ 4 insulated containers (Styrofoam cups with lids or aluminum foil, or other similar containers)
- ❒ 4 medicine droppers
- ❒ 16 oz. cottage-cheese-style container (to use as a waste container)
- ❒ 2 or 3 paper towels
- ❒ cafeteria tray

These density-related activities were adapted from the GEMS unit, Discovering Density. *That unit would make an excellent complement to* Ocean Currents. *You could present the density unit first, then reinforce and widen student understanding through the currents unit. Or, you could come up with other ways to combine the two units. In addition to a sequence of hands-on minds-on activities, the GEMS density unit teaches students the equation for the determination of density, and challenges students to apply density to a number of real-life scenarios, from heating systems and fishing to refrigerators and root beer floats!*

- ❐ stir stick
- ❐ 2 sets of colored markers or pencils (red, blue, yellow, green)

For each pair of students:

- ❐ 2 or more Density Layers Plan student sheets, page 82
- ❐ 1 Density Layers data sheet, page 83
- ❐ pencil

For the Demonstrating Density teacher demonstration:

- ❐ 4 (10 oz.) wide-mouthed, clear plastic cups
- ❐ tablespoon
- ❐ stir stick
- ❐ pouring container with 3–4 cups of water
- ❐ small bag of marbles
- ❐ box of salt (kosher, pickling, or table)
- ❐ cafeteria tray
- ❐ graduated cylinder or other measuring device

Optional

- ❐ spring or balance scale

Getting Ready

Preparing the Solutions

1. Write the following information on chart paper and label it Key to the Colors. Tape it up on a wall where everyone can see it.

red—hot and salty
blue—cold and salty
yellow—hot and fresh
green—cold and fresh

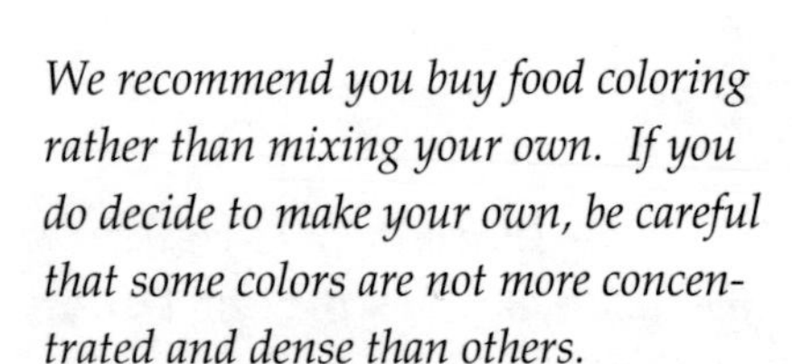
We recommend you buy food coloring rather than mixing your own. If you do decide to make your own, be careful that some colors are not more concentrated and dense than others.

2. Label the four thermoses as in step 1 above.

3. Add about 30 drops of the appropriate food coloring to the thermos labeled with that color.

4. Add the designated amount of kosher salt to the appropriate container:

red—14 level tablespoons (about ¾ cup) salt
blue—14 level tablespoons salt
yellow—no salt
green—no salt

If you use pickling or table salt instead of the coarser kosher salt, use 10 tablespoons (about ½ cup) instead of 14.

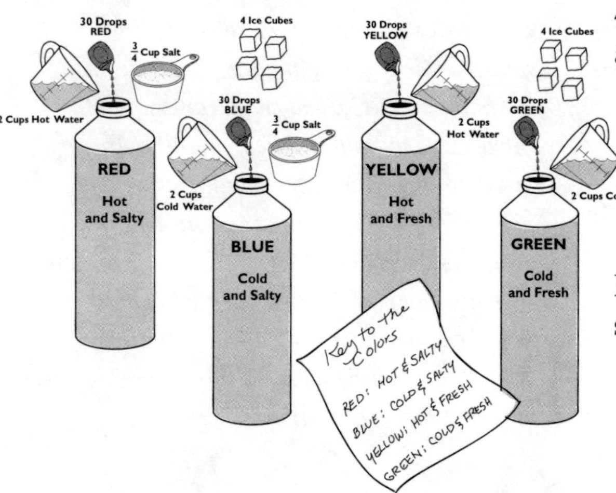

5. If ice cubes are not available at your school, you can bring them in an ice chest from home and freeze water in liter-sized plastic bottles to keep your cold source cold. Measure and add 2 cups of room temperature or cold water and 4 ice cubes each to the thermoses labeled:

blue—cold and salty

green—cold and fresh

6. Close the two cold water thermoses to prevent warming, and shake the blue one vigorously for about 30 seconds to dissolve the salt in the water.

7. Heat up about 6 cups of water. Determine how you will make and keep hot water in your room. If your classroom doesn't have really hot water, it works well to boil tap water in an electric kettle or coffee maker or use a stove or hot plate. Measure and add 2 cups of hot water each to the thermoses labeled:

red—hot and salty

yellow—hot and fresh

8. Close the two hot water thermoses immediately to prevent cooling, and shake the red one vigorously for about 30 seconds to dissolve the salt in the water.

9. Shake the blue thermos again for about 30 seconds to finish dissolving the salt in the cold water.

Prepare the materials for each working table of 4–6 students

1. Cut the raw potatoes into 1" thick slices. (You'll need two slices per pair of students and about five extras).

2. Cut two straws in half for each pair of students. Make some extras too.

3. Duplicate one copy of the Density Layers Plan (page 82) for each student and one copy of the Density Layers data sheet (page 83) for each pair of students. You may need more copies of the Density Layers Plan if students revise their plans.

4. Label four insulated cups for each table:

blue

red

yellow

green

One Inch Slices

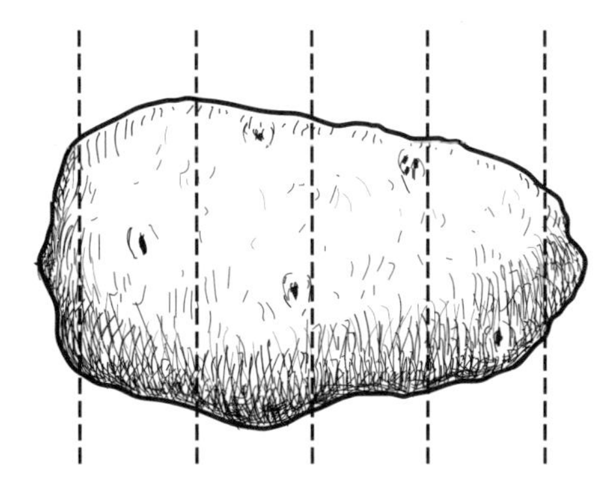

If you have uniform-sized marbles, you could be exact and weigh them, but it is not necessary for this demonstration.

5. Prepare one equipment tray per group of four to six students and place one medicine dropper in each insulated cup.

Other Preparation Steps

1. Prepare the Demonstrating Density teacher demonstration materials and place on a tray.
    a. Use a graduated cylinder or other measuring device to fill two of the cups with as close as possible to the same amount of water (about three-quarters full).
    b. Fill two other clear plastic cups about three-quarters full of marbles as equally as possible.
    c. Set these materials aside until later.
2. Make a small amount (about a tablespoon) of purple water for demonstrating the procedure by adding red and blue drops to water.
3. Arrange the tables or desks so two or three pairs of students may work together at one station. Be sure students' work surfaces are as level as possible.
4. Set all the student equipment trays, the extra straws, purple water, oil, and thermoses in a central location. **Be sure to finish the teacher demonstration before you distribute the student equipment trays.**
5. Write out the Key Concepts on chart paper.

- **The ocean is made up of layers of waters of different densities.**
- **Cold water is denser than warm water.**
- **Water with salt is denser than fresh water.**
- **The more closely packed the molecules in a substance, the denser the substance.**

Session 1: Introducing the Challenge

1. Tell students their challenge is to try to create layers of colored liquid by using only water and salt. Ask them what they learned about salt water compared with fresh water in the previous session. If they don't mention it, remind them the salt water formed a layer or water mass below the fresh water. Ask them what they remember about cold water versus hot water. If necessary, remind them the cold water formed a layer or water mass below the hot water.

2. They will now attempt to layer four colored liquids in the same straw so none of the colored liquids mix. Show them the equipment, and challenge them to make as many layers as they can. Suggest they first try to make two layers, then try for three, or even four.

3. Show them the liquids and the Key to the Colors chart you prepared.

4. Refer to the Density Layers data sheet and show them that the key to the colors is repeated there. Demonstrate how to color in and label their cylinder/straws with colored markers showing both their prediction and the results of their layering trials. Have them circle the sequences in which the liquids layered successfully.

5. Tell them, before they make their predictions and attempt the layering trials, they must first make a plan. Distribute the Density Layers Plan student sheet to each student. They will each make an individual plan and then compare their plan with their partner's plan. Together they will decide what they will do as a team. The student sheet also asks them to give a rationale for their proposed plan.

6. Tell the students if their initial plan appears to be unsuccessful, they are free to make a second or third plan—but they must write it down on an additional student sheet before trying it.

The liquids don't need to be added in the order of the most dense to least dense because they will slip and slide past each other to form stable layers. However, this will only work as long as the liquids are added slowly, gently, and at a 45° angle. You will probably want the students to discover this for themselves.

Demonstrating the Procedure

1. Caution the students not to taste any of the liquids.

2. Insert a straw into a slice of potato at a 45° angle. Explain that the reason the straw is placed at this angle is so the liquid can run down the inside of the straw. If the straw is pushed in too far, liquid may leak out of the bottom. If that happens, insert the straw into a different spot on the potato and try again.

3. If some of your students have no experience using medicine droppers, demonstrate the technique.

   a. Squeeze the bulb to expel as much of the air and contents as possible. Hold your squeeze as you insert the tip of the dropper into the liquid.

   b. Release your squeeze on the bulb so the liquid is drawn up into the dropper.

   c. Carefully squeeze the bulb to expel the liquid out into the straw, one drop at a time.

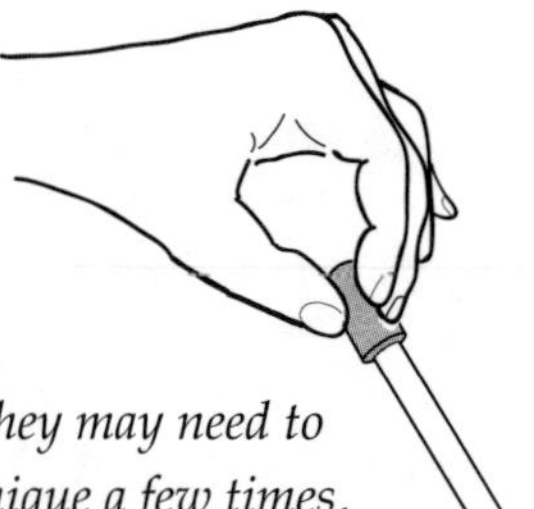

Remind students they may need to practice their technique a few times, or perhaps repeat a trial or two, in order to get the technique perfected so the liquids don't mix.

4. To demonstrate, gently add some purple water, drop-by-drop, down the inside of the straw, until about one-half inch of liquid accumulates. Tell your students you are using different liquids for the demonstration than they will use in their tests. Let them know the purple liquid is room temperature water. Then add some oil to the straw in the same manner. Point out that, when students do this, they should put in about a finger's width (½") of liquid so all four liquids can fit in the straw.

5. Hold a piece of white paper behind the straw so students can see the layering easily. Tell the students they can use their data sheets as a white background.

6. Show how to empty the contents of the straw by removing it from the potato over the waste container, and allowing the liquid to spill into the container. If a piece of potato plugs up the bottom of the straw, students can gently squeeze out the plug or, if necessary, discard the straw and get a new one.

7. Show students where the four thermoses are, and tell them they can pour a *small* amount from each into their four cups. Tell them, if they are using hot water, they should get a small amount in their insulated cup just before using it, so it will still be hot. Caution students not to mix liquids in the different cups and to keep medicine droppers in the same cups so as not to contaminate the solutions.

8. Before letting your students begin, make sure they understand their challenge: to layer two or more liquids in the straw so none mix together. Tell them to predict what will happen each time, and record their prediction on the data sheet.

9. Briefly demonstrate "sloppy" methods to show what the results would look like if the layers *do* mix.

10. Demonstrate how, after each test, partners draw and label the sequences they tried on their data sheet. Have them circle the sequences in which the liquids layered successfully.

Testing

1. Divide the class into teams of four to six students per work area. Distribute one equipment tray to each work area. At each table have students divide further into teams of two.

If you have time, you might want teams to make presentations of their plan and trials.

2. Distribute the Density Layers Plan student sheet to each student. After they have completed their plan, have them work together in pairs to decide on their team plan.

3. After the partners have completed their team plan, distribute the Density Layers data sheet to each pair. Have the students begin the layering trials. Circulate around the room, lending encouragement or assistance as needed. You may need to remind the students to make and record their predictions before actually trying their layering trial. Also remind them to circle the results of their successful trials.

4. When most of the students have successfully layered at least two liquids, have the teams place their materials on the trays, and have one student from each table return the tray to the equipment area.

Discussing Results

1. Tell your students to pin or tape their data sheets up on a designated wall or chalkboard space. Remove materials, clean up, seat your students, and draw their attention to the data. Ask them for general observations. If they don't mention it themselves, ask them if there is any color that shows up at the bottom of the cylinder drawings most often (probably the blue—cold and salty water). Ask them why they think this is. [It is the most dense.]

Be sure your students understand that color is just a marker for temperature and salinity.

2. Ask them if there is a color that frequently was the top layer (probably the yellow—hot fresh water). Ask them why they think that is the case. [It is the least dense.]

There is the possibility you may have to explain a discrepant result your students might get in one of their trials (e.g., hot, salty water forming the bottom layer).

How might you or your students explain this? Some possible explanations:

- *The water may have cooled off so it was no longer hot, but only room temperature*
- *Some unobserved mixing had occurred*
- *There might have been inaccurate preparation, handling or labeling of the solutions*

How could you test an explanation?

- *Redo the experiment and preparation of mixtures*
- *Use a thermometer to measure temperature and a hydrometer to measure density*

3. If any groups were successful at making four layers, draw the attention of the class to their data sheet. Ask them to rank them from least dense to most dense. [Hot and fresh (yellow), cold and fresh (green), hot and salty (red), cold and salty (blue).] If no one was able to make four layers, ask your class to help rank them in the order they think they would layer in.

4. Ask the students why the liquids formed separate layers. [The amount of salt in the solution, temperature differences, and method of adding them gently.]

5. How might they make even more layers? (Hint: Where would room temperature water fit in? What about using more salt?) What would layering of cold water or hot fresh water and room temperature salty water look like? If you have time, have the students try to make these layers. What would affect the results? [It would depend on how cold, how hot, or how salty the solutions were as to what the layers would look like.]

6. Ask students to think of situations involving the layering of liquids in the ocean. [Ideas will vary, but should be focused on the notion that the ocean is full of layers of waters of different densities, largely due to salinity and temperature.]

Density describes how concentrated matter is. A very concentrated drink (such as Kool-Aid without enough water added) has a lot of mix concentrated in a relatively small amount of water. Dense things have a lot of material concentrated in a relatively small space.

Session 2: Discussing Density

Salt Water and Fresh Water: Demonstrating Density

1. Show your students the two cups of water, and tell them you measured the same amount of water into each.

2. Show them the two cups of marbles, and tell the students to imagine the cups represent the two cups of water on a molecular (small particles) level. The marbles represent water molecules in the two cups—although there would actually be billions of water molecules in a cup of water.

3. Tell your students density has to do with how tightly packed molecules are together. A more dense liquid has tightly packed molecules, and a less dense liquid has looser packed molecules. Remind your students substances can have high densities if the molecules are simply more massive without being more closely packed.

Density = its mass divided by its volume

Density is a property used to describe all types of matter including liquids, solids, and gases

If you started with marbles of equal weight in each cup, and if you have access to a scale or balance, let students observe the difference in weight associated with greater density.

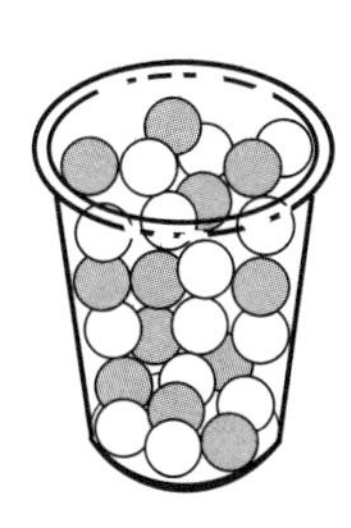

4. Now have them imagine that each of the two marble cups has the exact same amount of water, like the cups of water. Ask which would be the densest. [They would have equal density.]

5. Add two tablespoons of salt to one of the cups of water. While you stir it, tell the students this was how their salt solutions were made. Review with the students which of these liquids is more dense: the cup with only water, or the cup with water and salt? [The one with salt because more molecules are packed into the same container.]

6. Tell them they'll now see how this looks in the molecule model of the water glass. Pour some salt into one of the cups with the marbles. Ask what they observe. [The salt packs in around the marbles, filling in the spaces between the marbles.]

7. Explain that salt and water molecules mix in a comparable way, making the mixture denser. Ask the students which cup weighs more (the one with marbles and salt or the one with just marbles). [Salt cup weighs more.]

Introducing Temperature and Density

1. Tell them it's easy to see how, if something like salt is added, water can become denser—but what about water without other ingredients added? Ask them if they have an explanation for why cold water would be more dense than hot water.

2. Tell them all molecules, including those in liquids, are always moving. Point out that, as molecules move, they bang into each other and push each other away. This causes molecules to be spread out and take up more space—that is, fill a larger volume.

3. Let them know the hotter the liquid, the faster the molecules move, the more they hit each other, the more space between them, and that makes the liquid less dense. Make sure they understand the hot water still has the same amount of molecules in it, but the molecules are moving faster, and are spread out more.

Kinetic theory is difficult for students to grasp at this age level. Although they should understand that cold water is more dense than warm water, do not expect them to fully understand why after this brief explanation.

4. Explain that molecules in cold water move less, and so are more tightly packed together, making the liquid more dense.

Density and Currents

You may want to explain to your students that when salt water freezes the resulting ice tends to be less salty and the liquid water around the ice gets saltier.

1. Ask your students to brainstorm examples of locations where these four types of water might be found in the ocean. Examples you may choose to bring up, if your students don't, are included in brackets.

- Cold and salty. [Antarctica and the Arctic—water freezes, leaving the remaining liquid water very salty and cold.]
- Cold and fresh. [Canada and Siberia—places where northern cold rivers drain into the ocean.]
- Warm and salty. [Mediterranean Sea and Red Sea—the water is warm, but due to evaporation of water and low river inflow, it's also very salty.]
- Warm and fresh. [Amazon River and Congo River—places where warm rivers drain into the ocean.]

You may want to mention that plankton, fish, sleeping elephant seals, dead whales, waterlogged driftwood, dissolved substances, pollutants, and other things move with these water masses.

2. Explain that after these waters enter the ocean, they become ocean water masses that form layers which may last a long time and move long distances. For example, some of the dense Mediterranean intermediate water flows into the South Atlantic basin—and some of the cold, salty water that formed near Greenland may flow into the South Atlantic, Indian, and Pacific basins.

3. Tell your students their job was to make **stable** water mass layers—but in the ocean, these waters of different densities move, causing currents. Ask if any of them witnessed a "mini-current" in their straws, as one color moved below or above another. Ask them how waters of different densities might affect ocean currents.

Density in ocean water is determined by both salinity and temperature, acting together. Salinity is sometimes the dominant factor, but temperature is usually dominant because it varies so much more than salinity.

4. Lead a class discussion of these ideas:

- Warm water rising and cold water sinking can cause currents (like the cold water currents that formed in the Waste Disposal activity)
- Saltier water sinking and less salty water rising can also cause currents

5. Hold up the Key Concepts for this activity, and ask one or more students to read them out loud. Post the concepts on the wall for students to revisit during the rest of the unit.

- **The ocean is made up of layers of waters of different densities.**
- **Cold water is denser than warm water.**
- **Water with salt is denser than fresh water.**
- **The more closely packed the molecules in a substance, the denser the substance.**

Going Further

1. If your students are curious about whether a large quantity of a less dense liquid would still layer on top of a small quantity of a more dense liquid, you could test it. After making predictions, measure 5 ml of water with food coloring in a cylinder. Fill the rest of the cylinder with oil, and see what happens. Many students are surprised that, even with a large quantity, the less dense oil remains on top.

2. You could ask your students how could we know if something has a high or low density. One answer is to see if it floats or not. A warm water layer will float on top of a cold water layer. Fresh water will float on top of salty water.

DENSITY LAYERS PLAN

Color the diagrams showing the layer as follows:

red = hot and salty

yellow = hot and fresh

blue = cold and salty

green = cold and fresh

For my **first** layer I'll add ______________________________

Reasoning ______________________________

____

____

For my **second** layer I'll add ______________________________

Reasoning ______________________________

____

____

For my **third** layer I'll add ______________________________

Reasoning ______________________________

____

____

For my **fourth** layer I'll add ______________________________

Reasoning ______________________________

____

____

red = hot and salty
yellow = hot and fresh

DENSITY LAYERS

blue = cold and salty
green = cold and fresh

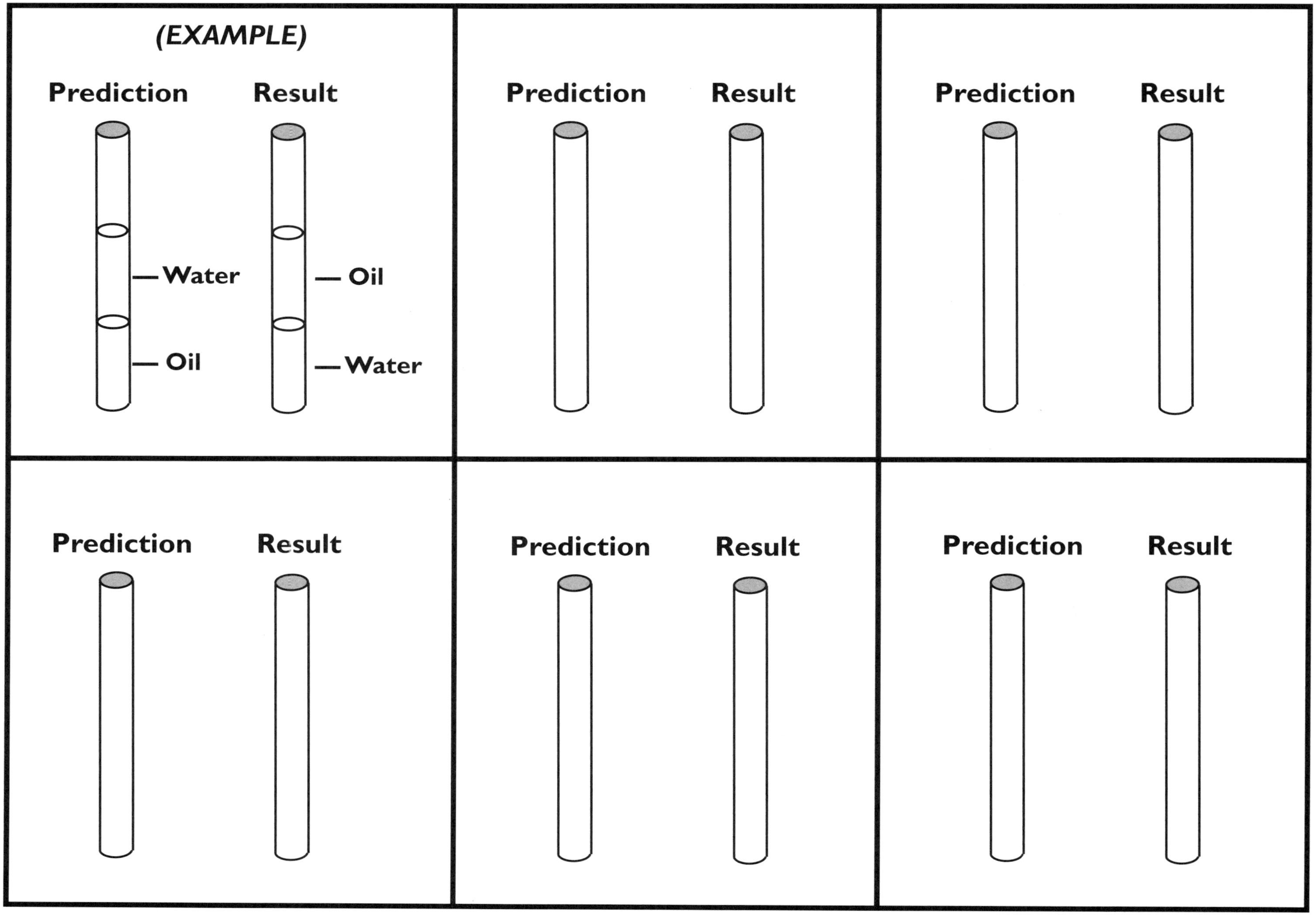

(EXAMPLE)

Prediction
— Water
— Oil

Result
— Oil
— Water

Prediction

Result

Prediction

Result

Prediction

Result

Prediction

Result

Prediction

Result

Activity 5: Ice Cubes Demonstration

Overview

The ice cubes demonstration synthesizes what students have learned about density-related currents. In this case, temperature and salinity are combined to look at the interactions that create ocean currents. Students make predictions about whether ice cubes will melt faster in fresh water or salt water and explain their reasoning. They watch an experiment and hypothesize about the results. The teacher checks for understanding as the students participate in a Think, Pair, Share activity to debrief the results and conclusions of Activities 1–5.

The purposes of this session:

- Give students a chance to observe and consider a phenomena that may be surprising to them (sometimes called a "discrepant event")
- Reinforce previous learning about the relationship of temperature, salinity, and density to the formation of ocean currents
- Connect their learning to the density of water near Antarctica

"This activity was awesome! I loved the discrepant event aspect, and it all came together here. I am pleased and surprised at how much knowledge they have gained."

"This really brought home the point of densities and currents to my students. That in itself was great, but I also have to admit that I really enjoyed seeing the puzzled look the students got when their predictions did not match up with their observations."

What You Need

For the class:

- ❒ 2 identical plastic jars (1½–2 liter clear plastic soda bottles with the tops cut off; plastic 18 oz. peanut butter jars; or wide-mouth mayonnaise jars); or beakers
- ❒ tap water (enough to fill jars)
- ❒ vial (¼ fl. oz) of food coloring (any dark color, not yellow)
- ❒ ¼ cup salt (kosher preferred)
- ❒ 8–10 ice cubes
- ❒ spoon
- ❒ 32 Think, Pair, Share student sheets, page 93
- ❒ 32 pens or pencils
- ❒ 10 sheets of chart paper
- ❒ 32 sheets of white paper
- ❒ masking tape

Kosher salt works best as the water won't appear so cloudy.

Getting Ready

1. Fill the two identical jars about ¾ full with tap water. Add about ¼ cup of kosher salt to one of the jars, mix thoroughly, and let sit.

2. If possible, have these charts from previous activities taped up or displayed in some way in the classroom.

- Key Concepts for Activities 1–4
- Student responses to the question posed in the Thought Swap in Activity 2: Waste Disposal—Describe some of the ways you can think of to make a current in a pool, tub, or even a glass of water?
- Student Posters from Activity 3: Current Trends

3. Write these questions on a sheet of chart paper and label it What Happened?

- In the station experiment with hot water and cold water, which temperature water ended up at the bottom?
- In the station experiment with the salty water and fresh water, which one ended up at the bottom?
- In which jar did the ice cubes melt fastest?

4. Make a copy of the Think, Pair, Share student sheet, page 93, for each student.

Making Predictions

Some teachers prefer to have small cooperative groups of students actually do this activity themselves rather than doing it as a teacher demonstration. This works very well, it is hands-on and the materials are easy to get. On the other hand, the teacher demonstration can be done in a very theatrical, entertaining, and dramatic way, which can be lots of fun and gives the students the opportunity to participate in a different kind of learning environment.

1. Show the students the two jars filled with water. Tell them they are the same except that one is salt water and the other fresh water, but don't tell them which is which.

2. Ask them how they could figure out which jar has salty water and which jar has fresh water. Distribute one sheet of paper per student and have them jot down at least two ideas (they'll write more on this sheet later). [Some of their ideas may include taste (not recommended and not a safe scientific procedure); put them in bottles with tornado tubes and food coloring; weigh them; or float something on them and compare how high the object floats.]

3. Discuss their ideas as a whole group, and then tell them that, although many of their suggestions would probably work, you could also tell which was which by using ice cubes.

Having students explain their reasoning can help turn a prediction into a more substantial hypothesis. Some teachers use a Think, Pair, Share discussion strategy to help students more fully articulate their reasoning.

4. Ask the students to predict whether ice cubes will melt faster in fresh water or salt water. Have them write down their prediction and explain their reasoning. Lead a class discussion about their reasoning, and then ask for a show of hands on their predictions.

Ice Cubes Demonstration

Some teachers said colored ice cubes work well–this way the melting can be observed from further away. But be careful as it might give away some of the story too early.

1. After the discussion of predictions, have a couple of students very carefully add 3–4 ice cubes to each jar (add the same amount to each). Be sure the ice cubes are of uniform size. You might even weigh them to ensure you are adding equal masses of ice.

2. Tell students to be especially careful to not bump or disturb the jars. (You might even consider leaving them undisturbed for a few days and watching what happens.)

3. Have each student draw and label the experimental setup and record their observations with diagrams and words on their sheet of paper used in Making Predictions.

4. Have a couple of students come and look in the jars. Ask them to tell the class in which jar the ice is melting faster. [Ice in fresh water melts faster.]

5. After the ice cubes have had an opportunity to melt some, carefully add 3–4 drops of food coloring to each of the jars, right on top of the ice, and instruct the class to watch what happens. Place a sheet of white paper behind the jars so the movement of the food coloring is evident. Ask students to record their observations.

6. Tell them this part of the exploration may help explain which ice cube melted faster. [In the fresh water, the food coloring disperses toward the bottom; in salt water the food coloring stays in a layer on the surface.]

7. Tape the What Happened? chart where everyone can see it. Ask the students to individually recall and write down the results of these experiments.

For English language learners, you may want to do this as a whole group discussion, referring to the station posters from Activity 3 still displayed around the room.

- In the station experiment with hot water and cold water, which temperature water ended up at the bottom? [Cold.]
- In the station experiment with the salty water and fresh water, which one ended up at the bottom? [Salty.]
- In which jar did the ice cubes melt fastest? [Fresh water.]

8. Lead a whole group discussion writing down their responses on the chart paper.

9. Let students know people often get confused about this demonstration, because they think about other experiences they've had with salt and ice. Tell them the results of this experiment **have to do with densities,** and not with freezing temperatures.

The predictions and hypotheses most people make regarding melting of ice cubes in salt water and fresh water are often based on their experiences using salt to make ice cream or salting icy roads. These experiences often lead them to wrong conclusions because they assume salt must cause ice to melt faster. Salt is used to keep roads from icing over because the salt lowers the freezing point of water. When making ice cream, salt again lowers the freezing point of water so the melted ice water is very cold and can remove more heat from the cream mixture over a longer period of time.

The key thing in the ice cube activity is that as the fresh water ice cubes melt, the cold water flows down and away from the ice cubes causing them to be constantly surrounded by warm water. As a result, the ice cubes melt faster.

10. Have students work in groups of four to come up with two or three basic statements explaining why the ice cubes melted faster in the fresh water. They should put their statements on chart paper. Remind them to consider all their observations and conclusions from the previous activities in the unit.

Discussing Explanations

1. Have each group of students present their statements and have the class analyze them. Lead a class discussion about their hypotheses.

- If your students were able to accurately explain the phenomenon, they may only require a brief synopsis of the following explanation.
- If your students are unclear, and you feel they could benefit, the following Guiding Questions may help their understanding.

Guiding Questions to Explain What Happened

1.) This experiment showed ice cubes melt faster in fresh water than in salt water. Describe what you saw happen in the fresh water jar. [In the fresh water jar, as the ice cubes melt, a current could be seen flowing toward the bottom as the icy water carries the food coloring down with it as it sinks. As the temperature evens out, the food coloring mixes throughout the jar.]

2.) Why did the icy water sink to the bottom in the fresh water jar? [The icy water is denser than the room temperature water and sinks rapidly to the bottom.]

3.) Why did the food coloring spread throughout the fresh water jar? [The sinking of the icy water helped set up currents in the jar which quickly mixed the icy water and the room temperature water. As the temperature evened out in the jar, there was no longer any temperature difference to act as a barrier to mixing. So the food coloring spread throughout.]

4.) Describe what you observed in the salt water jar. [The ice melted more slowly than in the fresh water jar, and the food coloring formed a layer at the surface.]

5.) Would you expect the surface, colored layer to be fresh or salty, and why? [Fresh because the ice cubes were made of fresh water. As they melted, the melted fresh water would float on top of the denser salt water.]

6.) Why didn't the icy cold water sink to the bottom in the salt water as it had in the fresh water jar? [The icy cold fresh water must be less dense than the salt water.]

7.) What kept the food coloring in the surface layer? How long would you expect the layers to remain the same? [The food coloring in the fresh water was trapped on the surface by the more dense salt water below. The layers should remain the same as long as the jar is left undisturbed.]

8.) Why did the ice cubes melt more slowly in the salt water jar? [The ice cubes melted very slowly in the salt water because they were surrounded by the icy just-melted fresh water floating on the denser salt water. It is as if the salt water jar had ice cubes melting in a refrigerator as compared to the fresh water jar which had ice cubes melting at room temperature.]

2. To connect these ideas to the ocean, explain that the densest water in the ocean is formed around Antarctica, because the water is very cold and very salty. This combination causes it to become very dense and sink to the bottom of the ocean basin surrounding Antarctica. This water then travels north as the densest water in the ocean.

There are many examples of salinity-related density currents, especially in the Arctic and Antarctic where very salty and cold water is created at the surface. It then sinks and flows around and away from the poles. *See illustration on page 92.*

Think, Pair, Share, or Revisiting Thought Swap

1. Tell students they are going to do a Think, Pair, Share activity to check their understanding of the concepts they explored in Activities 1–5.

2. Distribute a Think, Pair, Share student sheet to each student. Ask them these questions and have them first think about their answers, then jot them down on their paper.

- What is an ocean current? [Ocean currents are water in motion. They transport nutrients, pollutants, and heat throughout the ocean. They are made up of huge amounts of water moving fairly steadily in a fairly constant direction over fairly long distances.]

Participating in a Think, Pair, Share activity allows students some "think time" to formulate their own ideas and jot them down as notes. This gives all students, even those reluctant to answer questions, the opportunity to organize their thoughts before hearing the answer from a classmate or the teacher. This activity structure also allows the students to try out their ideas on a partner first in a "safe" setting, then together they can share with a small group, and finally with the class.

Think of a sinking water mass as gently and gradually sliding downward. A 1 km vertical descent may take place over a horizontal distance of 1000 km. This usually takes place over many years, so it's also very slow or sluggish.

- What sets water in motion and causes ocean currents to form? [Surface currents are caused by wind and density differences created by warming and cooling; input of fresh water, salts and other material; and water evaporation. Water that is more dense sinks and less dense water rises. In rising or sinking, there is almost always a lot of horizontal motion—as in the bottles connected by tornado tubes.]
- Describe some of the ways to make a current in a pool, tub, or even a glass of water.
- Why do you think ocean currents might be important? [Currents transport people, marine organisms, nutrients, and pollution around the globe; transfer heat from the tropics to higher latitudes; and affect the climate and weather. Downward currents transport oxygen and other gases toward the bottom; upward currents bring nutrients toward the surface.]

3. Have students share their responses with a partner, and then with a small group of four students. Tell them they can add to their student sheet if they hear something they agree with from another student.

One teacher said, "The students enjoyed being able to relate these results to the previous activities. They got excited at seeing little things fit into the big picture–this connected everything together. I loved this."

4. Lead a class discussion and record their ideas on chart paper. As appropriate, refer to the Key Concepts from the previous activities taped up around the room.

5. Refer back to the original chart paper from Activity 2 to compare what they now know and what they originally thought about "ways to make a current in a pool, tub, or even a glass of water."

What is the Key Concept?

1. Have small groups of students write **their own** Key Concept for this activity. Let them present their concept to the class.

2. Lead a class discussion comparing everyone's concepts and come to a class consensus on the wording of the concept(s). Write this on a sheet of chart paper and display it prominently.

3. Tell students they'll have a chance, in the next activity, to learn about explorers, past and present, who made use of ocean currents.

Going Further

1. Do a follow-up assessment with concept mapping. Provide a list of main ideas (e.g., the importance of currents, what is a current, causes of currents, role of density) and have the students make a concept map to connect these ideas together. You may want to post in the classroom all the main concepts addressed so far.

2. Make ice cream in class and use a thermometer to measure the temperature of the salty melting ice mixture. Compare this temperature to that of fresh melting ice water. (Salt water can be liquid at below 0°C.)

3. Have students do a parent and/or school survey including the following questions and others the class brainstorms.

- Why do you think salt is added to roads in the winter?
- Why is salt added to ice when making ice cream?
- Which will melt faster—ice in salt water or ice in fresh water?
- Why is salt added to boiling water?
- What is a current?
- What are all the ways to make a current in a glass of water?
- Describe how to make a current in a glass of water without moving it or the surface it is on.

4. Have the students repeat the melting ice demonstration at home for their parents. Have the parents first make a prediction and then diagram the results and/or write down their observations and conclusions.

5. Do research about where and how table salt is made and where in the world salt is used on roads. Also, you may want to explore problems caused by using salt on roads.

Antarctica is the coldest place on Earth because its huge expanse of year-round ice reflects most of the sunlight that strikes it. As cold sea water freezes, the surrounding water becomes saltier as the forming ice removes mainly fresh water from the ocean and excludes the salt (icebergs are 85% fresh water). So the sea water adjacent to the ice becomes both very cold and very salty. This very dense water sinks to the bottom of the ocean basin surrounding Antarctica and slowly works its way north along the ocean bottom. This water, known as Antarctic Bottom Water, may take more than a thousand years to reach the North Pacific basin, and approximately 750 years to reach 40° N latitude in the Atlantic (near Maine).

Think, Pair, Share

1. What is an ocean current?

2. What sets water in motion and causes ocean currents to form?

3. Describe some of the ways to make a current in a pool, tub, or even a glass of water.

4. Why do you think ocean currents might be important?

Activity 6: Ocean Routes

Overview

Students apply what they have learned about ocean currents to find the best routes for traveling across the ocean. They use information on wind-driven surface currents, density-driven deep currents, upwelling zones, and downwelling zones. Students go to different stations around the room at their own pace, drawing routes on their data sheet maps with colored pens.

During the summing up discussion, students share their ideas and routes. The teacher or small groups of students then show the actual routes on overhead transparencies while leading a class discussion about the voyages and situations.

This activity may take two sessions to complete all the scenarios. Some teachers have recommended saving some of the routes for homework.

The purposes of this session:

- Give students an opportunity to apply and transfer aspects of the hands-on and simulation-derived knowledge they have gained about ocean currents to real-life historical and maritime situations
- Provide students with examples of the ways ocean currents are and have been a major force throughout history—affecting human migration and the settling of people around the globe
- Encourage students to further investigate the prominent role ocean currents and the ocean played in human history and diverse world cultures

What You Need

For the class:

❒ 1 each of the 16 overhead transparencies, pages 109–124
❒ overhead projector
❒ colored pens or markers for the stations:
4 red, 4 blue, 2 green, and 2 orange
❒ transparency pens
❒ damp cloth to erase transparency pens
❒ 1 each of the station sheets 1–11, pages 125–135

For each student (or student pair):

❒ 1 each of the four student maps, pages 105–108

Getting Ready

1. Make copies of the four student maps, either one per student or one per pair, depending on how you'd prefer they work.

You could instead make a total of three overheads (A, B, and one overhead for overheads 1–11). In this case, you'd use erasable overhead pens to draw, and then erase, each of the voyages in turn.

2. Make one copy of each of these overhead transparencies:

A. Surface Currents
B. Deep Currents
1. Benjamin Franklin Atlantic Route
2A. Magellan: First Ship Around the World
2B. Captain Cook's Second Trip Around the World
3. Whaleship *Essex*
4. Nike Shoe Spill
5. *Endurance*
6A. Atlantic Swimmer Benoit Lecomte
6B. Atlantic Swimmer Guy Delage
7A. *Kon Tiki*
7B. *Ra I* and *Ra II*
8. Pacific Rower Gerard d'Aboville
9. Penguin Feather
10. Richest Fisheries of the World
11. Waste Dumping

3. Decide if you will save some of the scenarios as homework. If you plan on sending homework, decide which stations you will not set up around the room. Make copies of those stations to send home.

You may decide to do multiples of fewer stations, perhaps choosing those you think would be of most interest to your students.

With 11 stations around the room you'll be able to accommodate 22 students working in pairs. For large classes this means more than two students will be working at each station. If you think it may be a problem to have several students at a station, you could choose to have more than one of some stations.

4. Make copies of station sheets 1–11. Set them up as stations around the room, placing the designated colored pens at each.

1. Benjamin Franklin—red and blue pens
2. Traveling Around the World—green pen
3. The Whaleship *Essex*—orange pen
4. Nike Shoe Spill—red pen
5. The *Endurance*—orange pen
6. Swimming the Atlantic—blue pen
7. *Kon Tiki* and *Ra*—red pen
8. Rowing the Pacific—blue pen
9. Penguin Feather—green pen
10. Rich Fisheries—red pen
11. Waste Dumping—blue pen

Introducing the Stations

1. Ask your students to brainstorm real-life situations where knowledge of currents is important. Let them know that in this activity they will use information they've learned about ocean currents to figure out possible ocean routes. Then they'll learn what routes were actually taken.

2. Show Surface Currents on the overhead projector. Tell your students they will use this map during the activity. Remind them of the patterns of gyres. Ask them if any parts or symbols are confusing, and help explain them.

3. Show Deep Currents on the overhead projector. Remind students what upwelling zones, downwelling zones, and deep currents are. Tell them they will also use this map. Ask if any parts or symbols are confusing, and clarify as needed.

Downwelling zones are places where the surface water moves downward.

4. Explain to students that the routes they choose may not match the actual routes from real life, but they don't need to match. Students do, however, need to follow the patterns of the currents shown on the maps. Later, they will be shown the actual routes, but that doesn't mean the routes they predicted *couldn't* have happened.

5. To get the students started, you may want to use the Benjamin Franklin story as an example. Why might sailing ships that left England for the Colonies arrive faster if they traveled south to Northwest Africa rather than straight across to New York? Obviously, they were not traveling the shortest route, but they often arrived as much as two weeks sooner. What was going on? A second good example would be Richest Fisheries of the World. This scenario allows you to describe and review upwelling.

Looking only at the currents, ships of that time might also have sailed north to Greenland. However, the Greenland route was not used very much due to problems with storms and ice.

6. Assign students to their first station to make sure they are spread out evenly, and to make sure each station will be done by at least one person/pair.

7. Let students move through these stations at their own pace. Suggest they go to those they find most interesting first, since they may not have time to do all the stations.

8. Distribute the student maps, then allow your students to begin the stations.

9. You may choose to stop the activity when most or all of your students have completed the stations.

Instead of the teacher reading and describing all the scenarios, pairs of students could read the Station Debriefings for themselves and then present them to the class.

Debriefing the Stations

1. You may choose to go through the questions on the station sheets numerically, or you could start with those your students are most interested in.

2. Follow the same procedure for each station sheet.

- Ask a few students to share the route(s)/locations they chose. Compare these with the currents shown on the map. Use transparency pens on the Surface Currents and Deep Currents overheads, and/or let your students show their proposed routes.
- Point out any trouble spots in their routes that go against currents shown on the map. Let them know it is possible to sail, swim, row, or motor against a current, but that it is much more difficult and slower.
- Show the transparency with the actual route for comparison. Remind your students their route need not match the one shown, but that it does need to adhere to the patterns of currents shown on their maps. Compare the student's suggested route with the actual route by overlaying the two transparencies. Encourage your students to point out trouble spots they might have encountered, and reasons why they think another route might be better.
- Share any of the additional information listed in the Station Debriefings, and conduct a discussion as you deem appropriate to the interest level of your students. It is not necessary to discuss each one at length, but more information has been included in the Station Debriefings in case your students are interested.

Going Further

Have the students go to the library or on the Internet to research additional voyages including Columbus, the *Mayflower,* and the *Titanic.* See "Resources" on page 145 and "Literature Connections" on page 156 for books describing some of the exciting voyages in this activity as well as many others.

Station Debriefings

1. Benjamin Franklin

Transparency 1:
Benjamin Franklin
Atlantic Route, 1769

In 1771, Benjamin Franklin worked for the American Postal Service. He was asked to find out why it took English mail ships two weeks longer to get from England to America than for American ships to make the same trip. To find out, Franklin asked his cousin, a whaling captain from Nantucket. His cousin told him British ships sailed against the Gulf Stream current, sometimes moving farther backwards in a day than forwards. He said they had been advised to get out of the current, but British captains were "too wise to be counseled by simple American fishermen." Franklin and his cousin together drew up a chart of the Gulf Stream current which British captains then began using, shortening their travel time.

On later voyages he made across the Atlantic, Franklin used a thermometer—to tell if the ship was in the Gulf Stream or not. The Gulf Stream waters come from the south and are warmer; the waters from the north that are not part of the Gulf Stream are colder. If the water temperature around the ship became too cold, he knew his ship was no longer in the Gulf Stream.

2. Traveling Around the World

Transparency 2A:
Magellan: First Ship Around the World, 1519–1522

Transparency 2B:
Captain Cook: Second Trip Around the World, 1768–1771

Magellan (1480–1521) is generally credited as the first European explorer to circumnavigate the Earth (from 1519–1522). Notice that Magellan sailed through the middle of the Pacific on his way west. He would have been better sailing with the Peru Current along the west coast of South America, but he didn't know where the currents were and depended on the wind to power his sailing ship. Although he was killed in the Philippines before completing the voyage, one of the three ships he set out with completed the voyage.

Captain James Cook (1728–1779) made three major trips, twice sailing around the globe. The route he took on his first voyage was similar to Magellan's. On his second trip (1772–1775) he traveled in the opposite direction. His goal was to chart the Pacific, which is why his route looped in the Pacific. He was killed in Hawaii on his third major voyage, in 1779.

The assignment to propose a route around the world does not say in what year. Some students may propose routes that utilize canals that did not exist during the time of Magellan and Cook, and that's fine. The Magellan and Cook routes are provided as examples, not as right answers. Also, both Magellan and Cook were on sailing ships. They depended mostly on winds, but also used currents.

Transparency 3:
Whaleship *Essex*, 1820

3. The Whaleship *Essex*

The story of the *Essex* is one of the main inspirations for Herman Melville's famous novel, *Moby Dick*. Their story is a wild one. The crew started their voyage from Nantucket sailing in the Gulf Stream and stopped at the Azores (a group of islands off Portugal). It then took them five weeks of fighting against the fierce currents to get around Cape Horn at the southern tip of South America. After months of travel, killing many sperm whales and collecting many barrels of spermaceti, the twenty whalers (ages 15–29) were hunting a pod of whales in the mid-Pacific, and had killed three. (Sperm whales have a space in their head that is full of oil and spermaceti, which is a waxy substance. The whale uses this oil-filled space during echolocation. People at the time used the oil for lamp light.) During the hunt, one sperm whale slowly approached the ship, then speeded up and rammed it, moved away where it thrashed about in the water, then rammed the boat for a second time. The ship sank, leaving three lifeboats and the 20-man crew adrift.

There were many Quakers on Nantucket Island, who were as opposed to slavery as they were to war. They abolished slavery on the island in the early 1770s. Whaling crews included Native Americans, Quakers, African-Americans, and Portuguese. Highly respected as navigators, many African-Americans also became harpooners, mates, and at least two became captains. The Essex's *crew of 20 included six African-Americans.*

They had barrels of water to drink and hard bread and a few of the hundreds of giant tortoises they had caught in the Galapagos Islands to eat. The three boats managed to stay together for many days and nights as they drifted hoping to reach South America. One day, one of the boats was attacked by what was apparently a huge shark, which bit a chunk out of the boat. They fought it off with a pole and repaired the hole with some boards they had brought along.

After 30 days, the starving crew landed on a small island. Unfortunately, the island was too small to feed them all so only three men remained, and the rest moved on in the boats. (The three men later found eight skeletons from what was apparently a previous shipwrecked crew that starved to death there.) The boats eventually became separated at night, and one was never heard from again. The others drifted as the crew starved.

At first they gave sea burials to their dead, but then began eating them to survive. Even this was not enough on one boat, where they drew lots, shot and ate the loser, who died willingly. After about three months of drifting 3,500 miles, the five survivors on the two remaining boats eventually were rescued. One lifeboat was just to the north, and the other just to the south, of the Islas Juan Fernandez, off Chile. Another ship went back to save the three crew members on the island, who also survived.

4. Nike Shoe Spill

Transparency 4:
Nike Shoe Spill, 1990

Six months after the Nike shoe spill near Alaska the shoes started to wash ashore. Beachcombers made a party of it, and started organizing meetings to put together matching pairs of shoes, many of which had barnacles growing on them, but were still wearable. Here are the dates and locations of when and where the shoes washed ashore.

5/27/90	Nike shoe spill, North Pacific Ocean
12/90	Cape Flattery, Washington (200 shoes)
2/91	Vancouver Island, B.C., Canada (100 shoes)
3/91	Queen Charlotte Island, B.C. (250 shoes)
3/91	Washington coast (200 shoes)
4/91	Columbia River (350 shoes)
5/91	Oregon-California coast (200 shoes)
5/91	North Vancouver Island (200 shoes)
2/93	North coast of island of Hawaii (several shoes)

5. The *Endurance*

Transparency 5:
Endurance, 1914

Leaving his shipwrecked crew on an island near Antarctica, Ernest Shackleton and five men managed to cross the Antarctic Circumpolar Current in their tiny boat headed for the island of South Georgia off Argentina. Had they miscalculated in navigating, they could easily have missed the island, and been doomed. After 16 days and more than 700 miles on the howling seas, Shackleton and his crew landed on South Georgia. They knew there was a whaling station on the island, but unfortunately they landed on the opposite side.

The Antarctic Circumpolar Current is also called the West Wind Drift on some maps.

With three incapacitated sailors left behind in a cave, Shackleton and two others set out across the glaciers and mountain range to get to the other side of the island. After three days of agonizing travel, they found the whaling station. The workers there didn't believe their story. They couldn't believe anyone could have traveled from Antarctica in a 22-foot boat or even have hiked across the island, which no one had done before, nor would do again for many years. Eventually, they were convinced, and rescued the three men on the other side of the island.

Shackleton then tried three times to reach the crew on the island in Antarctica, but his ship couldn't make it through the ice. He went to Argentina to get a bigger ship. Finally, in August 1916, Shackleton made it through and found all 22 members of his crew, 9 months after the ship had sunk. Not a single person died, and there were no serious injuries!

Transparency 6A:
Atlantic Swimmer
Benoit Lecomte, 1998

Transparency 6B:
Atlantic Swimmer
Guy Delage, 1995

6. Swimming the Atlantic

Benoit Lecomte set out to swim across the North Atlantic Ocean to raise money for cancer research. He swam 6–8 hours a day and rested on a 12-meter yacht. Lecomte started at Hyannis, Massachusetts, and traveled more than 3,700 miles in 72 days to get to the French town of Quiberon. (You might want to ask your students if any of them have ever swam *one* mile, and how hard that was.) Lecomte swam with a Shark POD, which sends out an electronic field to keep sharks away.

Guy Delage chose a different route. He swam west from the Cape Verde Islands in Africa, and arrived at Barbados (southeast of Puerto Rico) 55 days later. He traveled 2,350 miles, swimming 6–8 hours a day, eating and sleeping on a 15-foot raft.

Transparency 7A:
Kon Tiki, 1947

7. *Kon Tiki* and *Ra*

Thor Heyerdahl was convinced Polynesians were so different from Southeast Asians they could not have been descended from Asian ancestors. He believed the first Polynesians had drifted—intentionally or accidentally—from South America. He noted the number of similarities in artistic styles between the Polynesians and native South Americans and that the sweet potato's Polynesian name is *kumara*, which is similar to the term *kumar* used by some of the Quechua people of Peru.

The name Kon Tiki *comes from the legend of the Sun King, who was said to have led his fair-skinned people toward the setting sun to Polynesia more than 1,500 years ago on balsa rafts.*

To demonstrate this theory, he and five companions cut down and lashed together South American balsa logs to form a raft which they named *Kon Tiki* and set out from a Peruvian seaport. After sailing and drifting aboard *Kon Tiki* for about 5,000 miles, a voyage that took 101 days, Heyerdahl and his friends landed on an atoll in the Tuamotu Archipelago east of Tahiti.

Heyerdahl did prove the trip was possible, but it's important to note that the *Kon Tiki* had to be towed 50 miles off shore to where the westward currents began. Modern anthropologists who have studied languages and genetics think Polynesians probably came from Southeast Asia.

But how could Southeast Asians have sailed to Polynesia, when the prevailing winds blow in the opposite direction? Some have proposed that they might have made the voyage during an El Niño year when the winds might have blown in the opposite direction.

In 1952, Heyerdahl elaborated on his theory, suggesting the Polynesians were descendants of Caucasoids from Bolivia, who were themselves perhaps of North African origin. In 1969 he set out in *Ra*, a reed boat made for him by boat builders from the African nation of Chad, using paintings of papyrus boats found in ancient Egyptian tombs as a guide. After about 3,000 miles, the boat had to be abandoned. In 1970, Heyerdahl had Indians from Bolivia, who still built reed boats, build *Ra II*. *Ra II* completed the 4,000 mile voyage from Morocco to Barbados in 57 days.

Transparency 7B:
Ra I (1969) and Ra II (1970)

Ra *is the name of the Egyptian sun god.*

8. Rowing the Pacific

Gerard d'Aboville left Choshi, Japan, and 134 days later arrived at a small fishing village in the state of Washington. He rowed 12 hours per day and crossed 6,300 miles of ocean. He endured two typhoons, blazing heat, and freezing cold. In 1980 he rowed across the Atlantic Ocean in 72 days.

Transparency 8:
Pacific Rower Gerard d'Aboville, 1991

Mick Bird wants to be the first man to row around the world solo! Also, in December 1999, Tori Murden became the first woman to row across the Atlantic Ocean. See "Resources" on page 145 for more information and Web sites for both these rowers.

9. Penguin Feather

This is not based on an actual event, and many answers are possible. Antarctic bottom water is the densest in the world ocean. As it sinks, it mixes with water from the Antarctic Circumpolar Current. Some of this water creeps up the bottom of the Pacific, taking about 1,000 years to reach the equator, and about 1,600 years to reach Alaska.

Some of the water also moves up the Atlantic ocean floor. It takes about 750 years to reach as far north as Maine.

Transparency 9:
Penguin Feather

10. Rich Fisheries

Upwelling zones are areas where cold water, rich with nutrients, comes from the depths up to the surface. The upwelling zones, which are good fishing areas, are marked on the map. Also marked are areas of rich fisheries where wide continental shelves, seasonally high light levels, and other factors create good fishing areas. Rich fisheries shown on the overhead are:

- The Gulf of Mexico because of the nutrients coming from rivers draining much of North America.
- The Alaska Coast because of cold water currents. Also during the summers, arctic areas have lots of nutrients and 24 hours of sunshine, so they are rich with life.
- Off Cape Cod because of small, localized upwelling areas.

Transparency 10:
Richest Fisheries of the World

Not all upward motions that bring nutrients closer to the surface are caused by wind and upwelling. Other processes can cause the upward motion—stormy and turbulent areas, areas of shallow water, and areas where masses of water meet, with one sinking and the other rising. These processes play an important role off much of Alaska, the northeast United States, and southeast Canada.

Transparency 11:
Waste Dumping

11. Waste Dumping

This is not based on an actual event, and many answers are possible.

The northern dump site is at a deep water downwelling zone. The waste would most likely flow along the bottom of the ocean in any direction the arrows point. The waste could return to the surface at any upwelling zone it encounters.

The equatorial dump site is an upwelling zone. The waste would most likely flow with the surface currents.

MAP 1
SURFACE CURRENTS

RED pen for Station 1
BLUE pen for Station 1
GREEN pen for Station 2
ORANGE pen for Station 3

MAP 2
SURFACE CURRENTS

RED pen for Station 4
ORANGE pen for Station 5
BLUE pen for Station 6

MAP 3

SURFACE CURRENTS

RED pen for Station 7
BLUE pen for Station 8

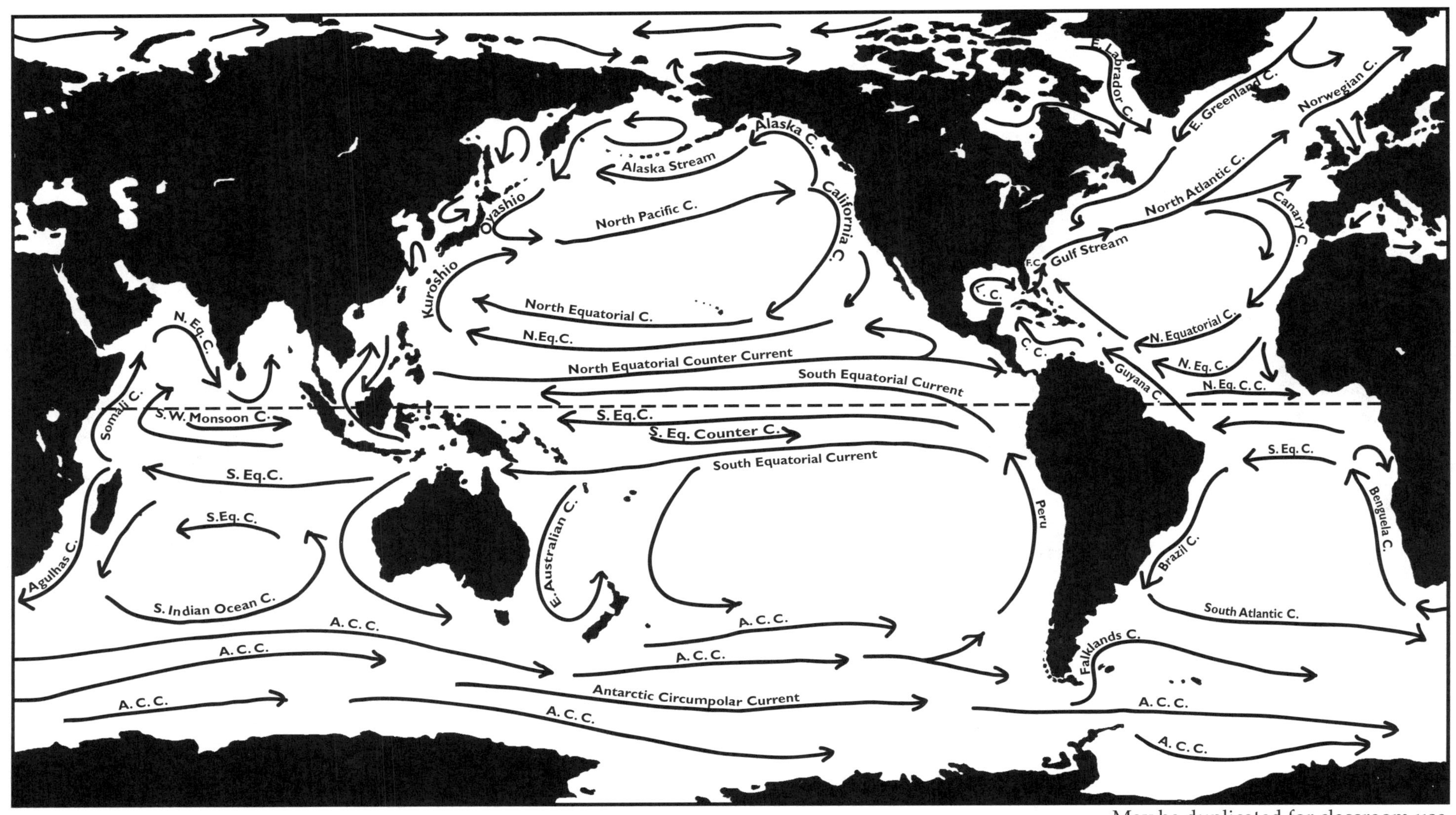

MAP 4
DEEP CURRENTS

GREEN pen for Station 9
RED pen for Station 10
BLUE pen for Station 11

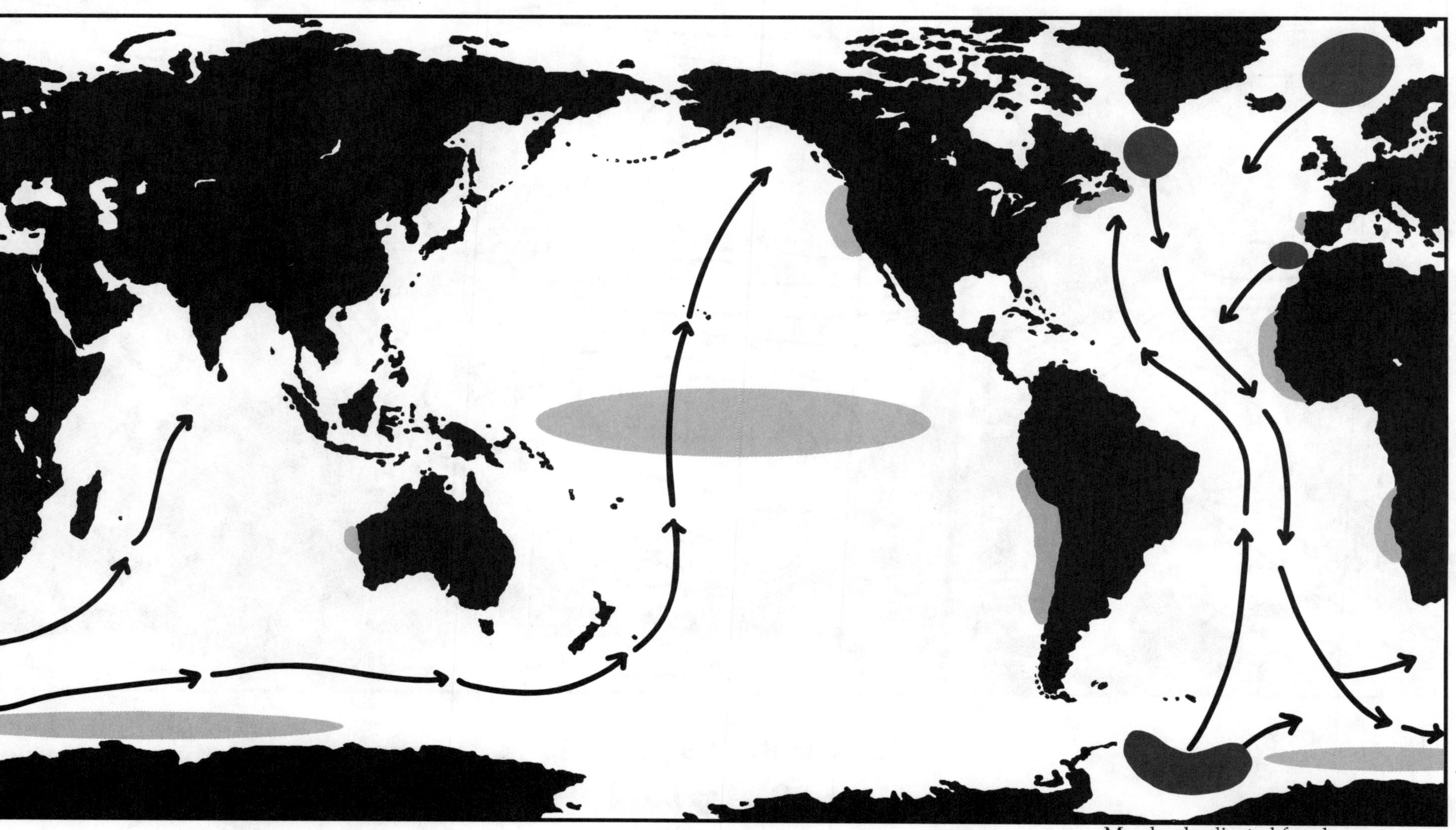

SURFACE CURRENTS (MAY–SEPTEMBER)

E. = East
S. = South
N. = North
W. = West
C. = Current

F. C. = Florida Current
L. C. = Loop Current
S. Eq. C. C. = South Equatorial Counter Current
C. C. = Caribbean Current
N. Eq. C. C. = North Equatorial Counter Current

S. Eq. C. = South Equatorial Current
A. C. C. = Antarctic Circumpolar Current
S.W. Monsoon C. = Southwest Monsoon Current
N. Eq. C. = North Equatorial Current

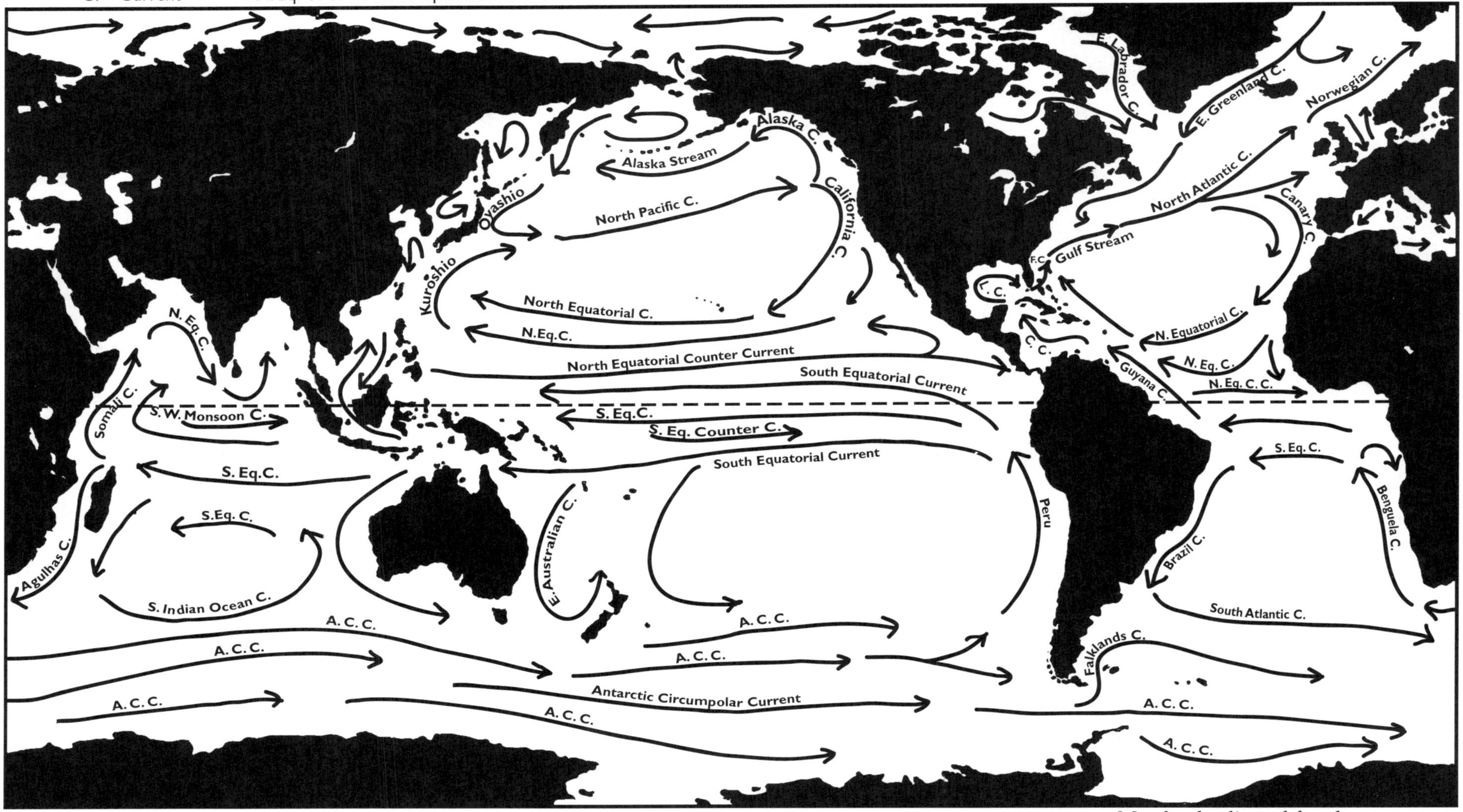

May be duplicated for classroom use.

DEEP CURRENTS

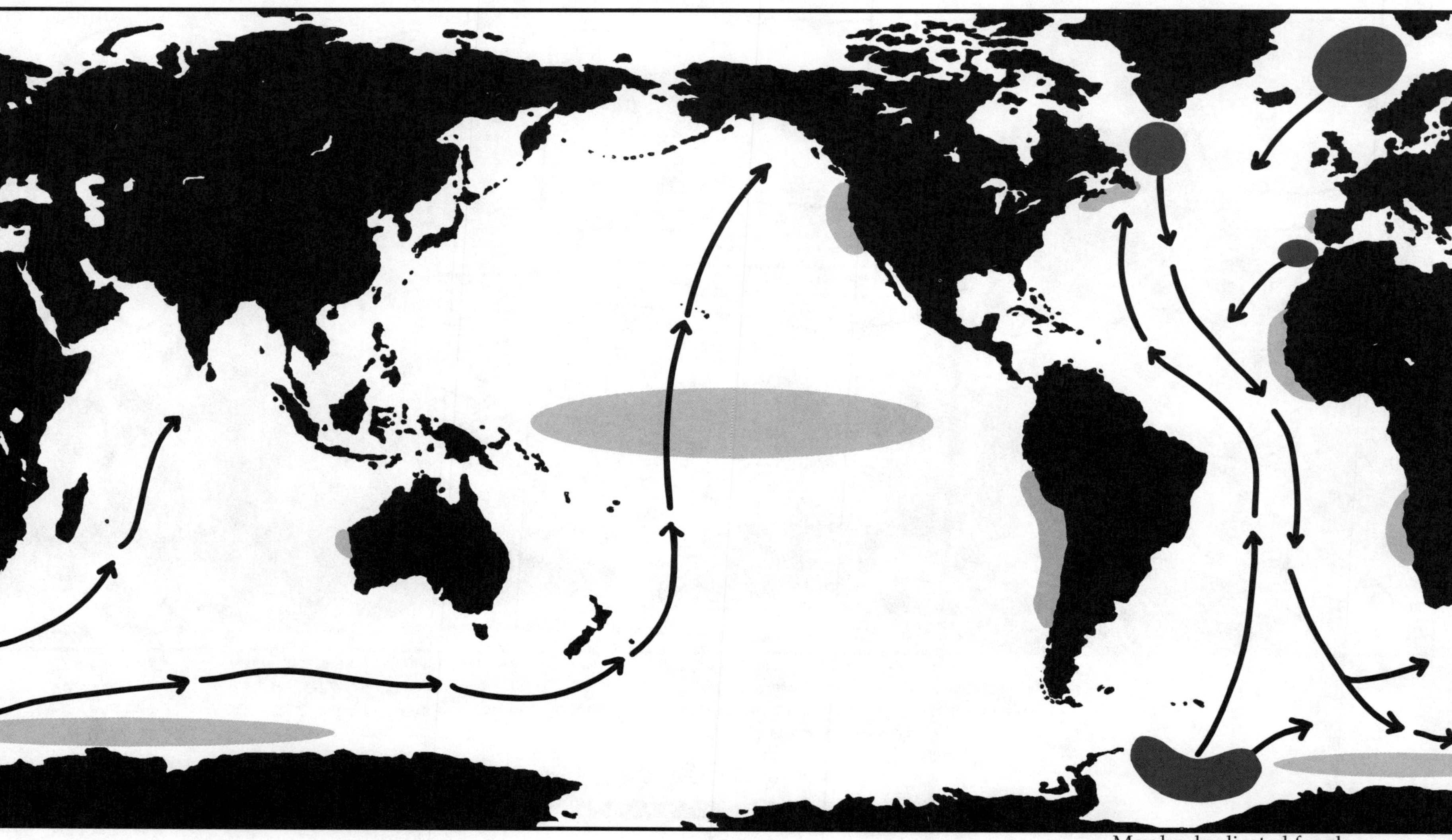

BENJAMIN FRANKLIN ATLANTIC ROUTE (1769)

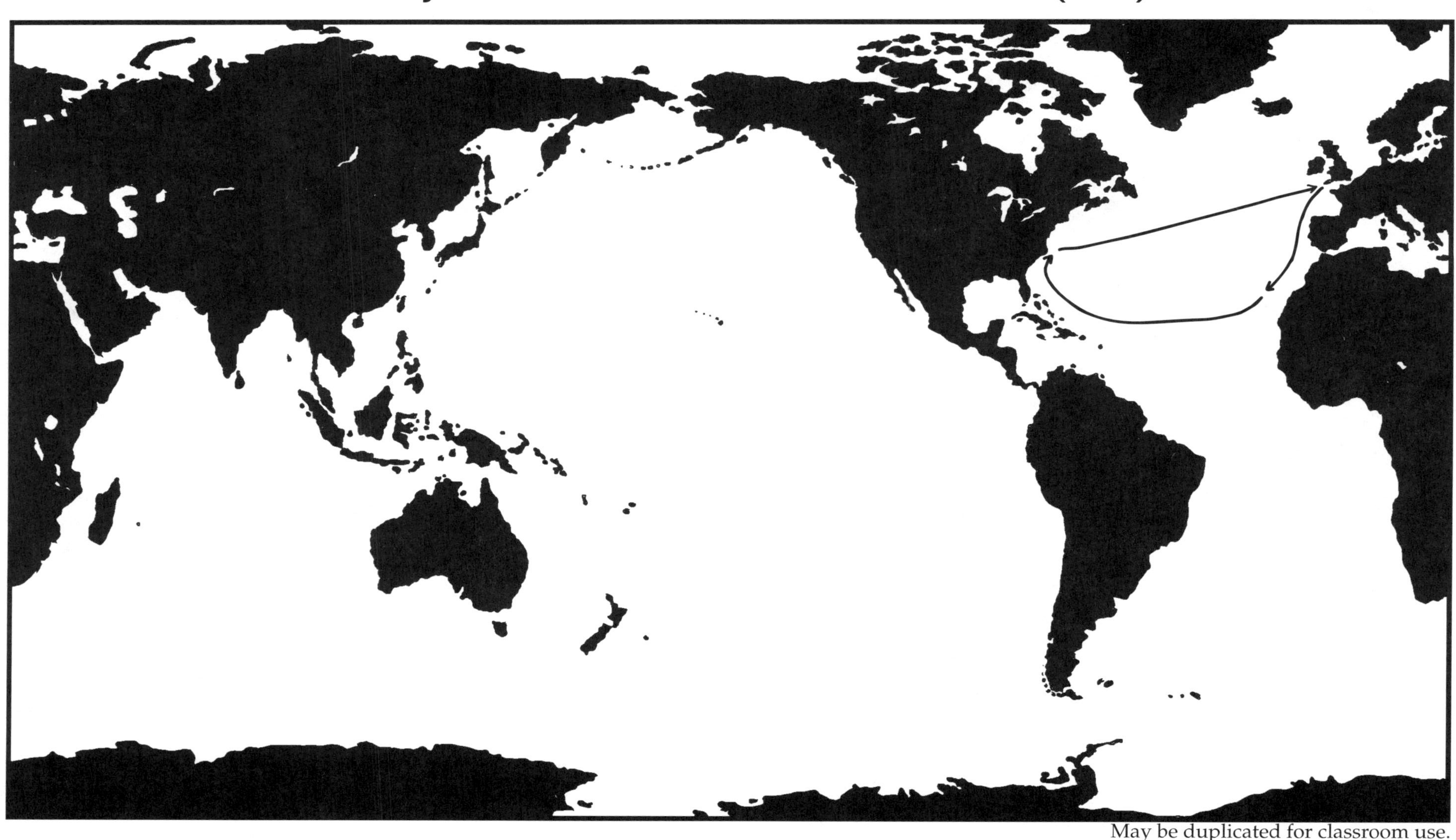

MAGELLAN: FIRST SHIP AROUND THE WORLD (1519–1522)

CAPTAIN COOK'S SECOND TRIP AROUND THE WORLD (1772–1775)

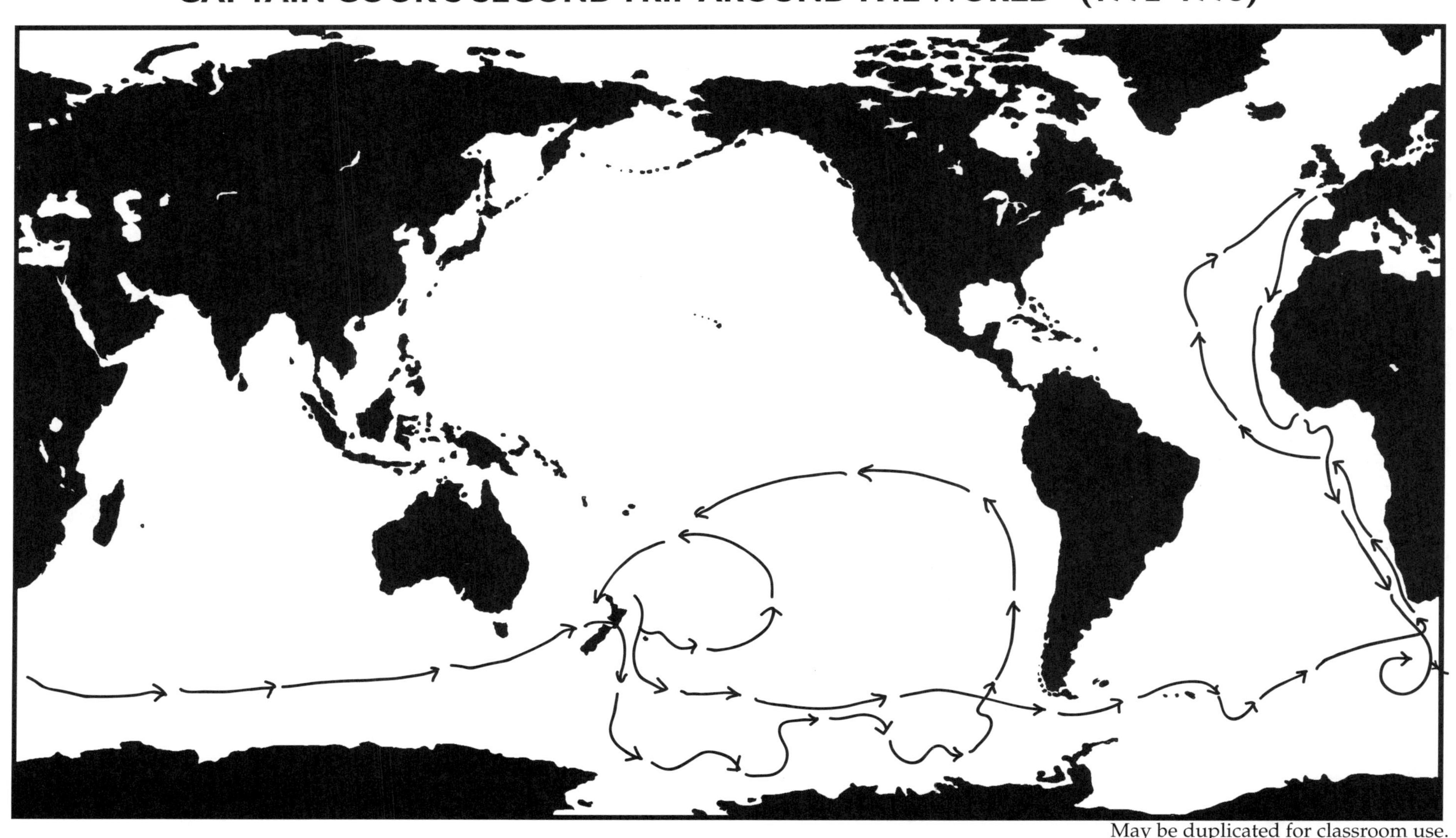

WHALESHIP *ESSEX* (1820)

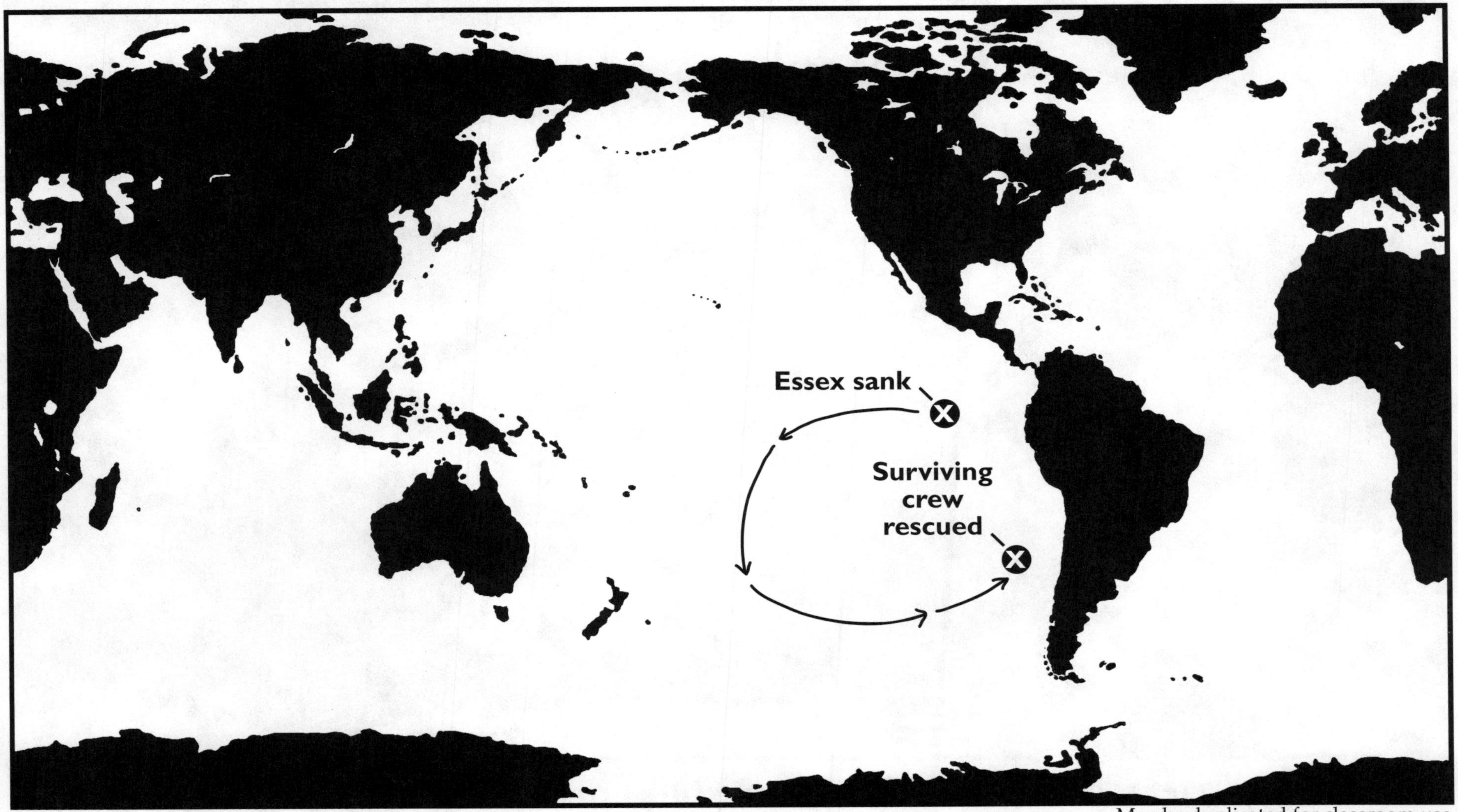

Essex sank

Surviving crew rescued

NIKE SHOE SPILL (1990)

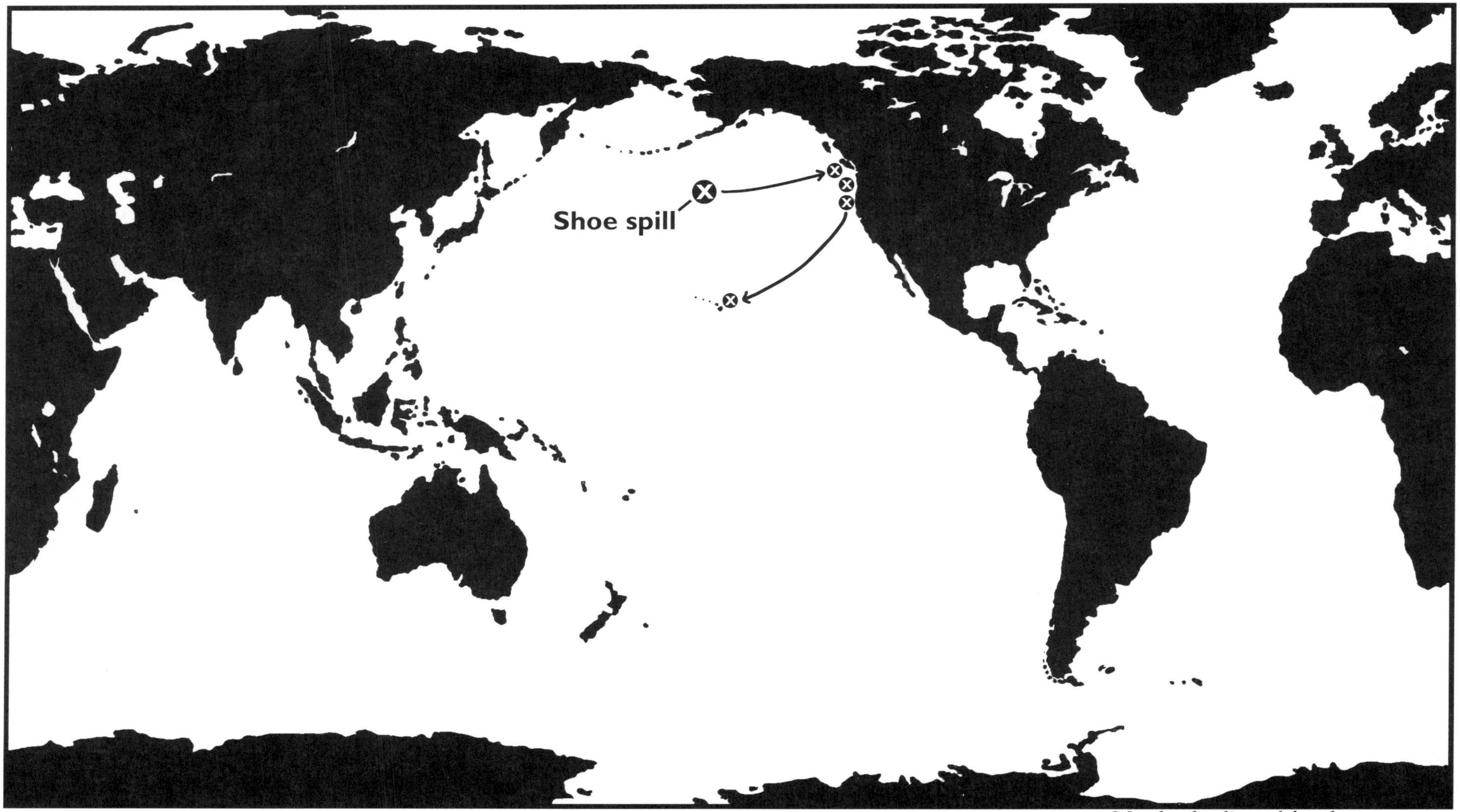

Shoe spill

ENDURANCE (1914)

ATLANTIC SWIMMER BENOIT LECOMTE (1998)

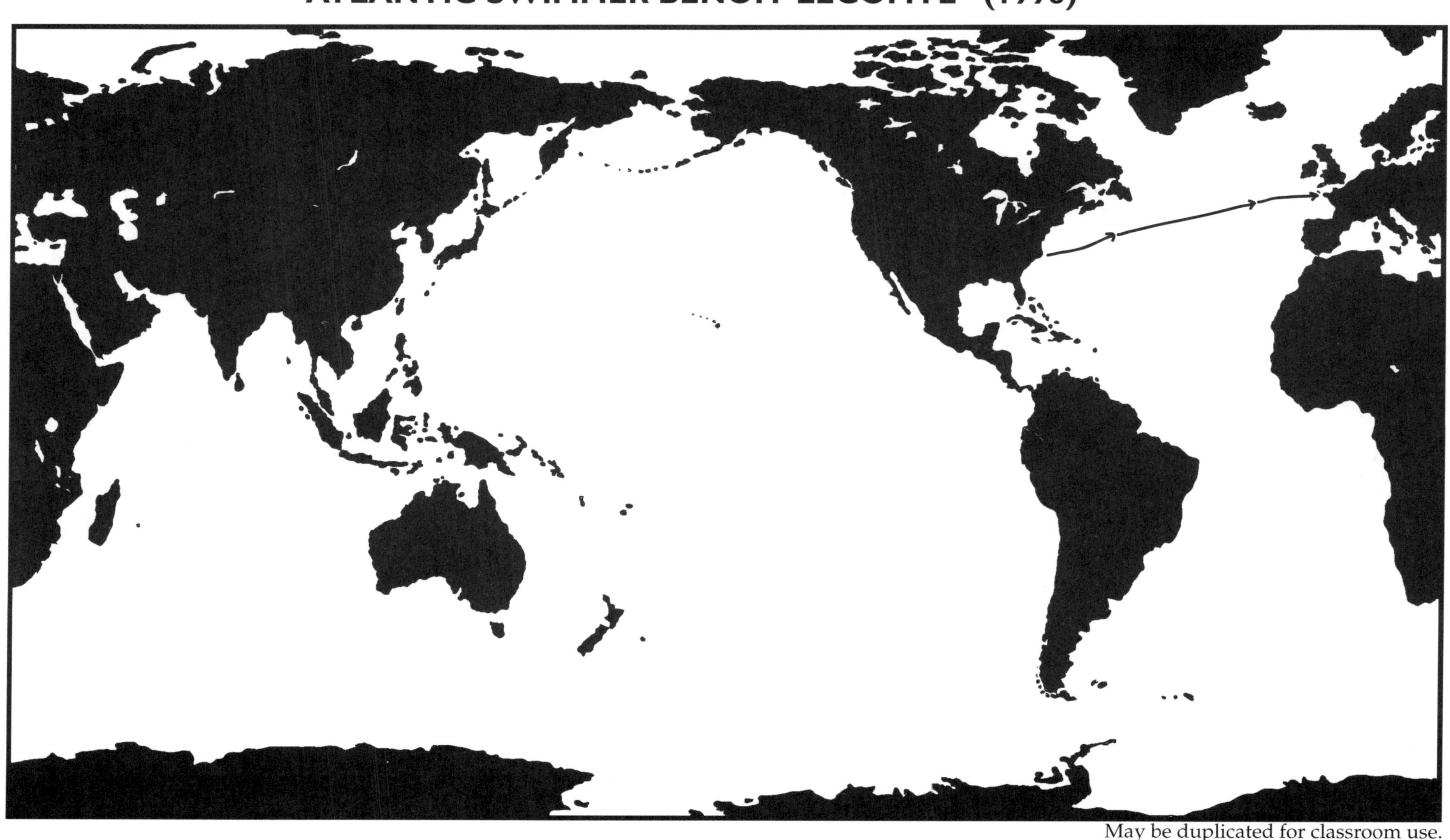

ATLANTIC SWIMMER GUY DELAGE (1995)

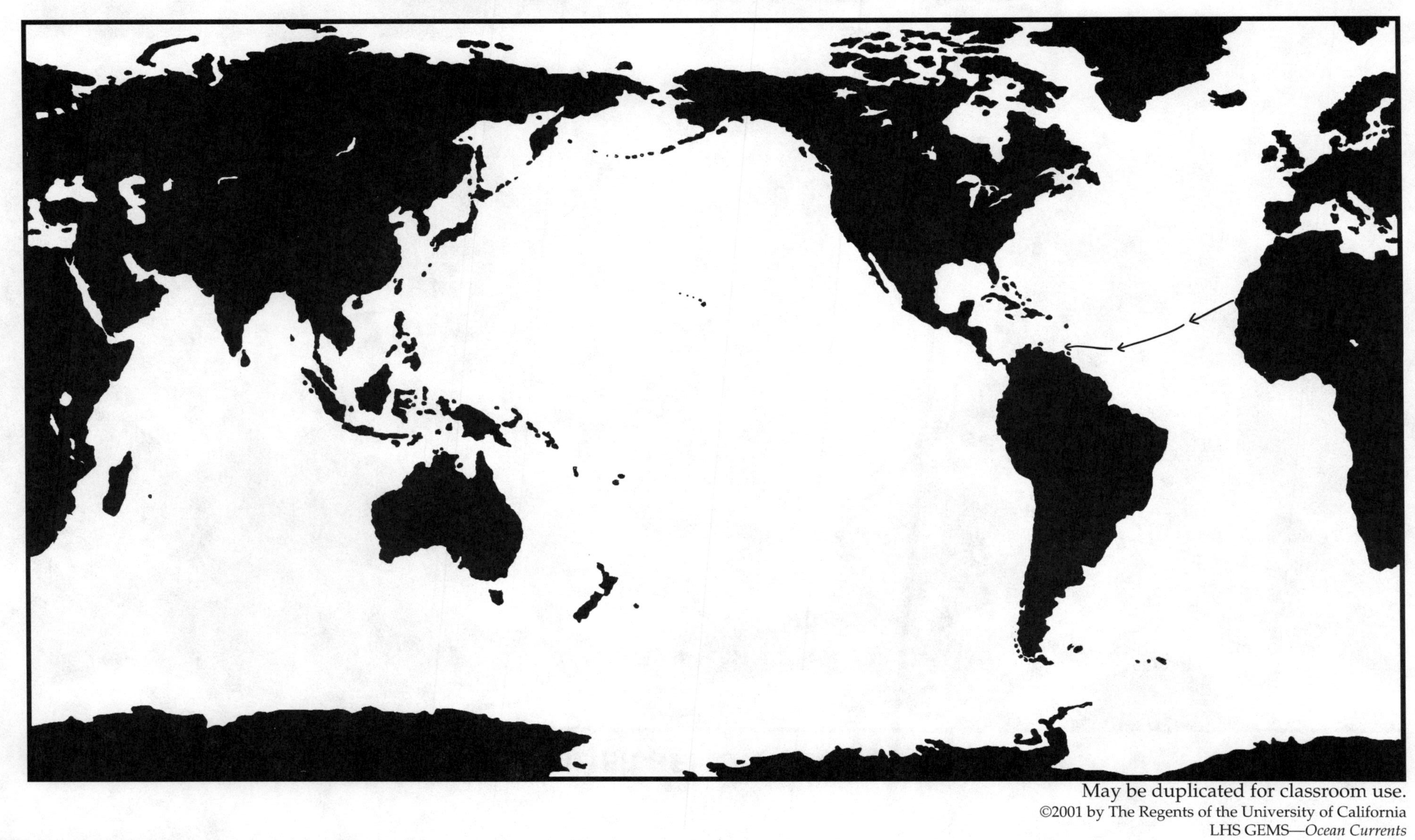

KON TIKI (1947)

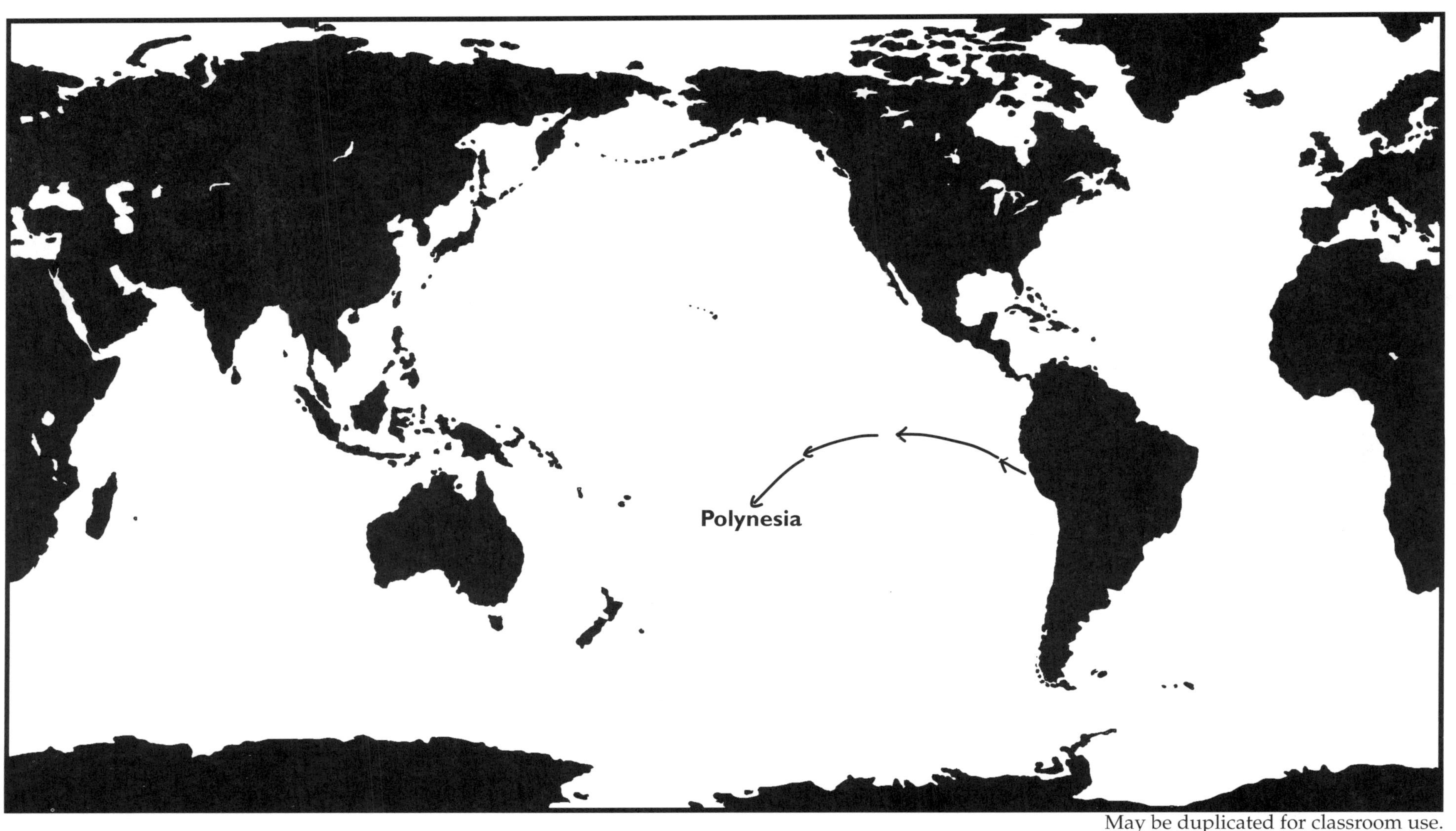

RA I (1969) and *RA II* (1970)

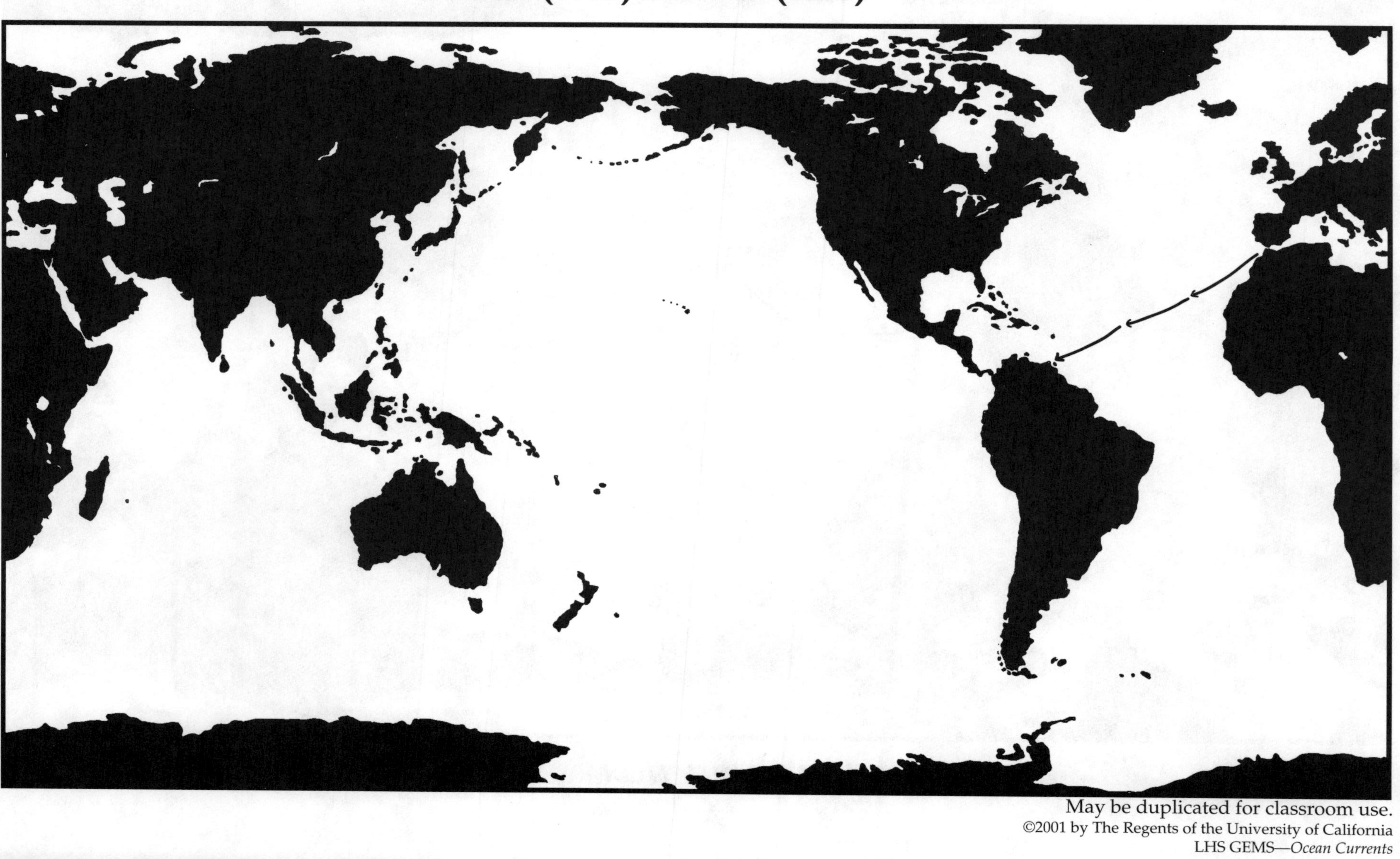

PACIFIC ROWER GERARD D'ABOVILLE (1991)

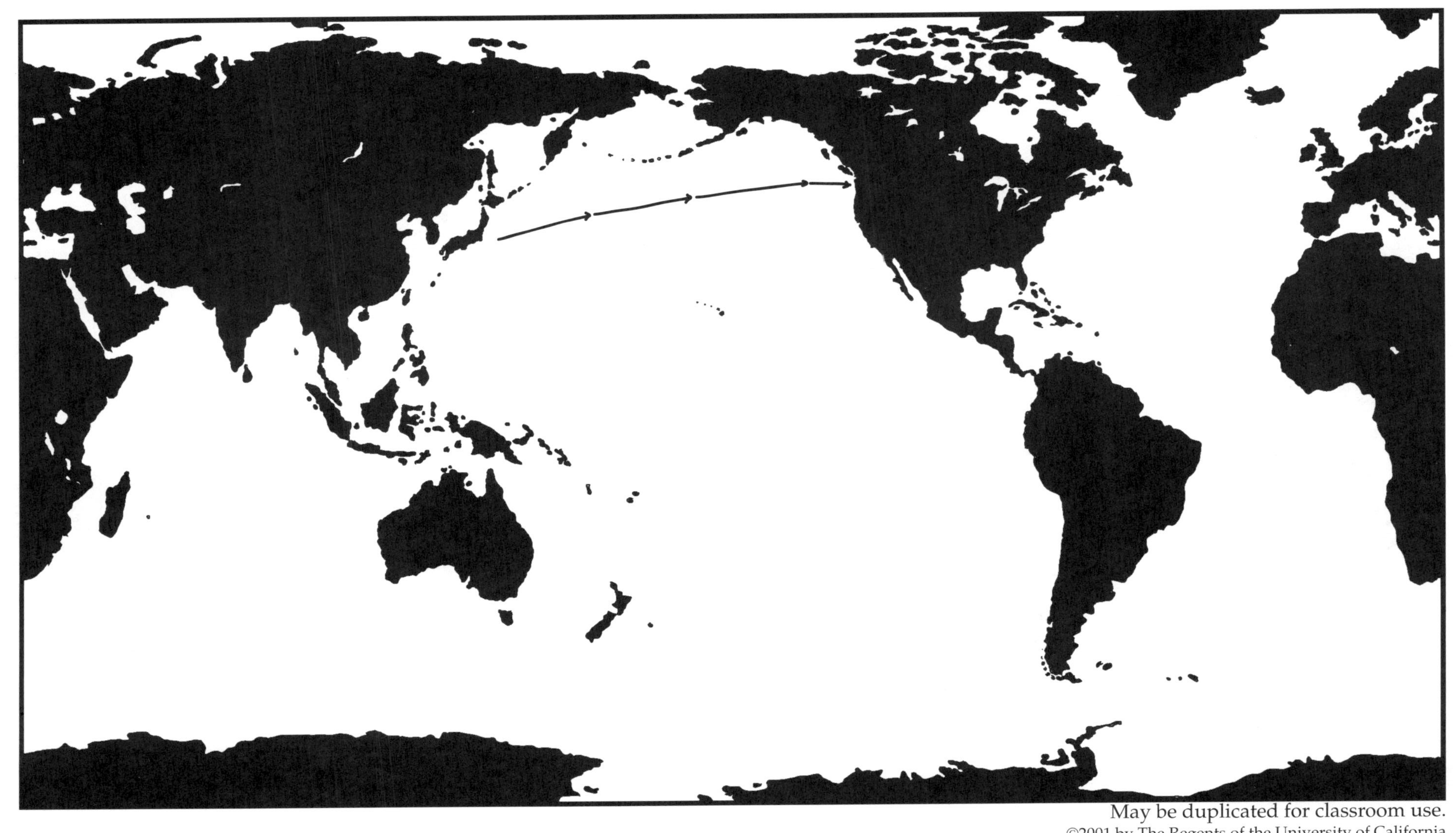

PENGUIN FEATHER

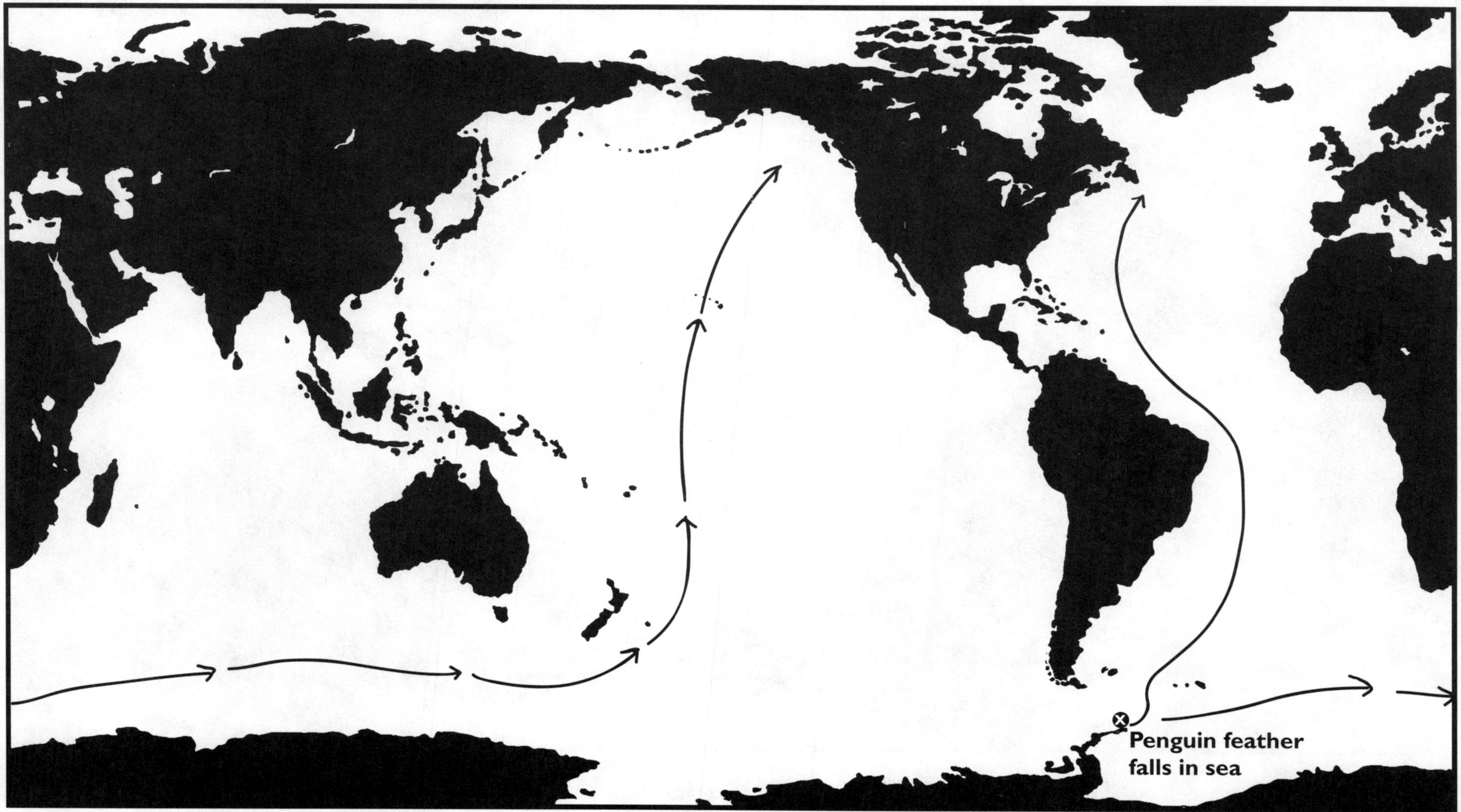

Penguin feather falls in sea

RICHEST FISHERIES OF THE WORLD

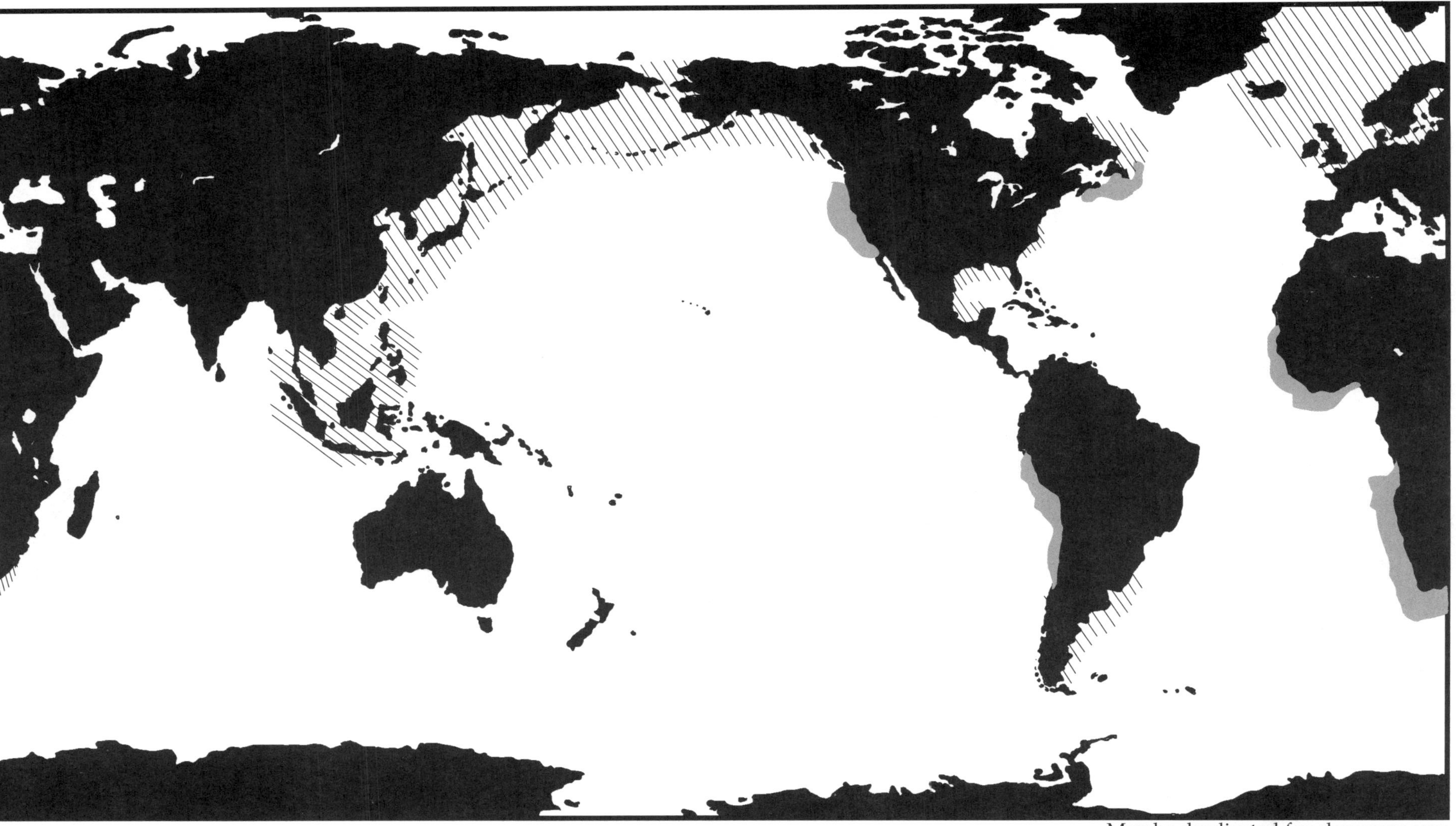

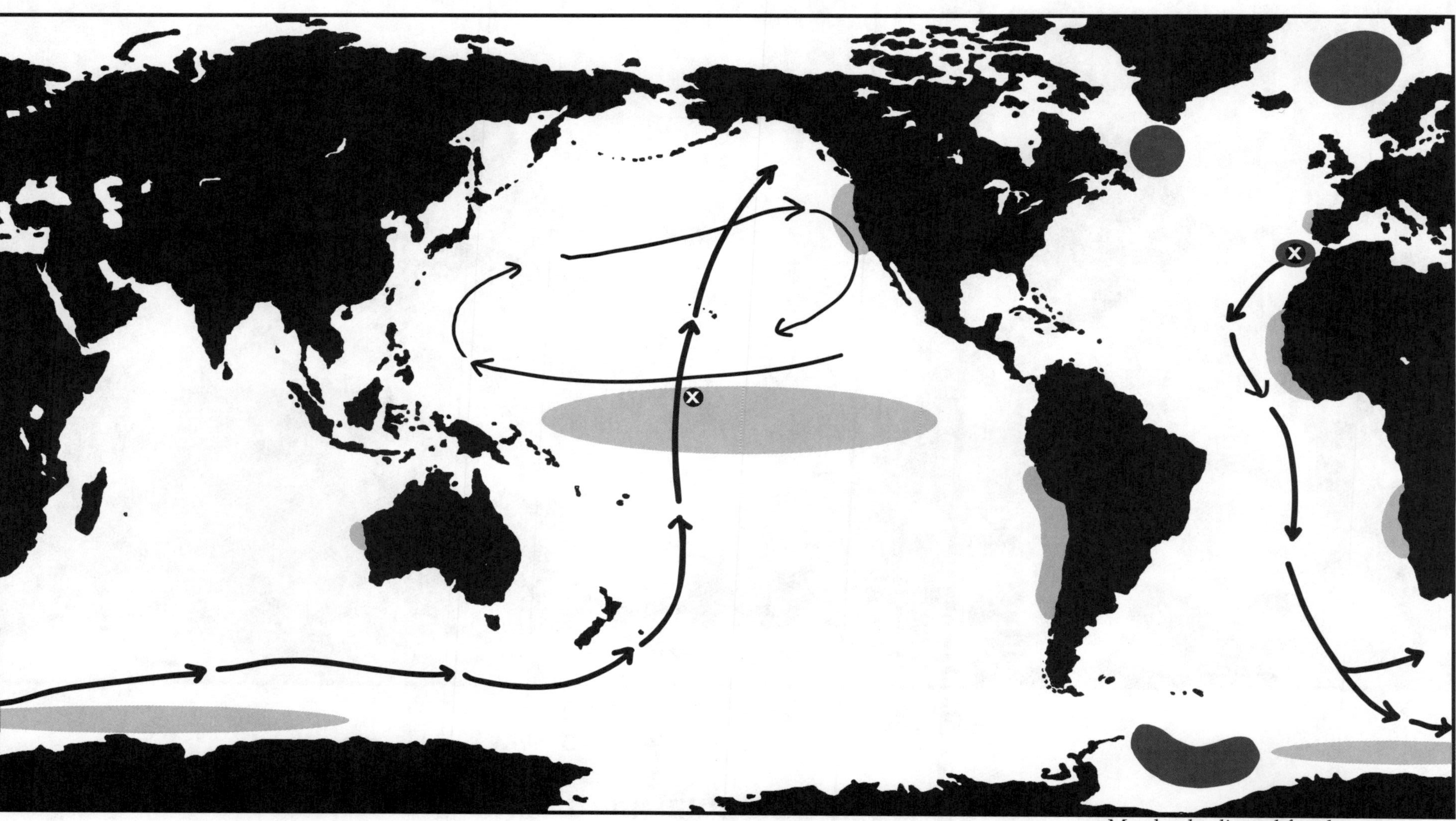
WASTE DUMPING
UPWELLING
DOWNWELLING

1. Benjamin Franklin

In 1771, Benjamin Franklin was asked to figure out why it took English mail ships two weeks longer to get from England to America than for American ships to make the same trip. Use a **RED** pen to mark the slow route you think the British ships were taking, and then with a **BLUE** pen, mark a route you think would be quicker.

Use **MAP 1** to mark your routes.

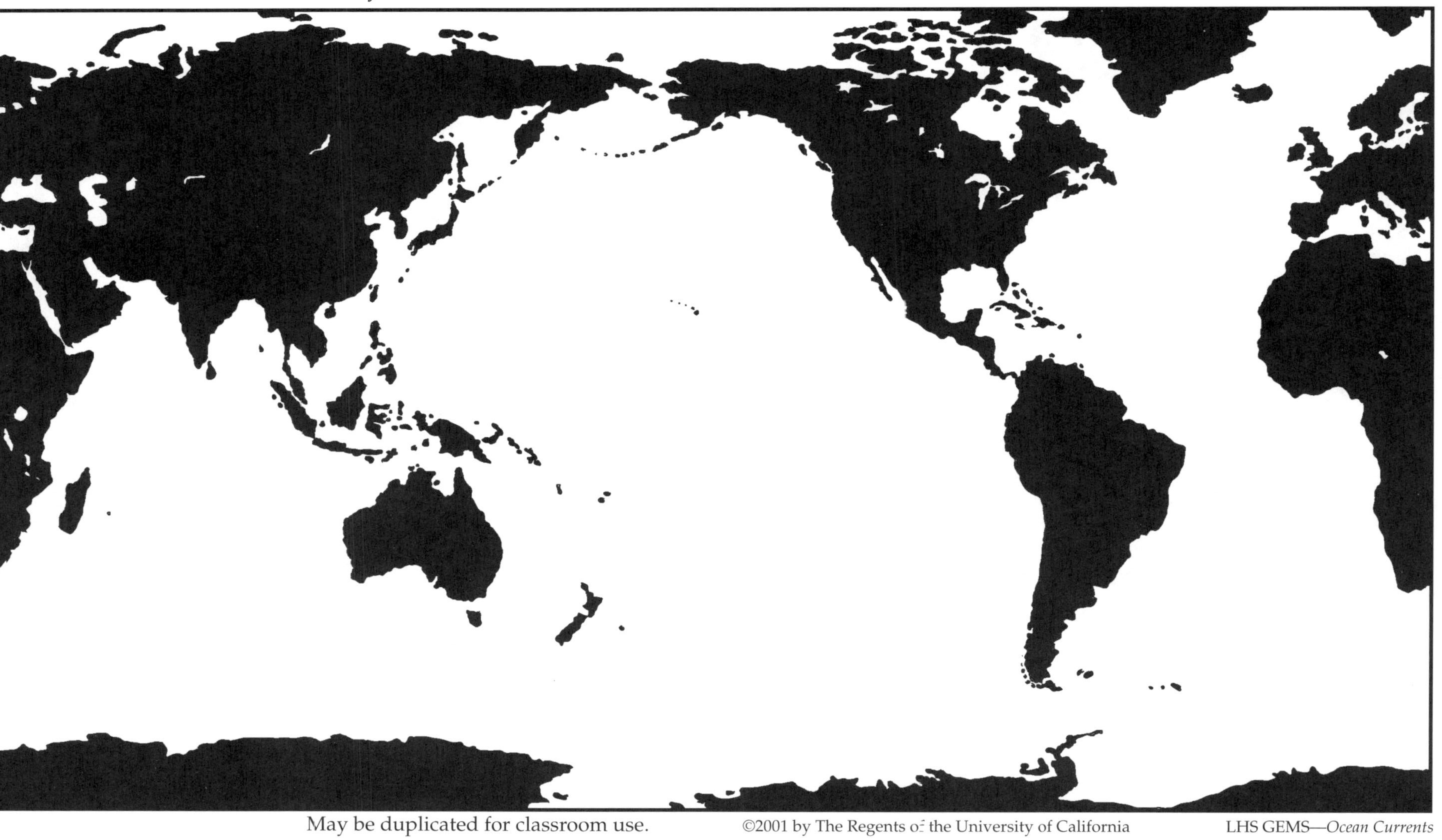

2. Traveling Around the World

If you were to travel around the world on a ship that depends on currents, mark with a **GREEN** pen what you think would be the easiest route. Your journey must begin and end in England.

Use **MAP 1** to mark your route.

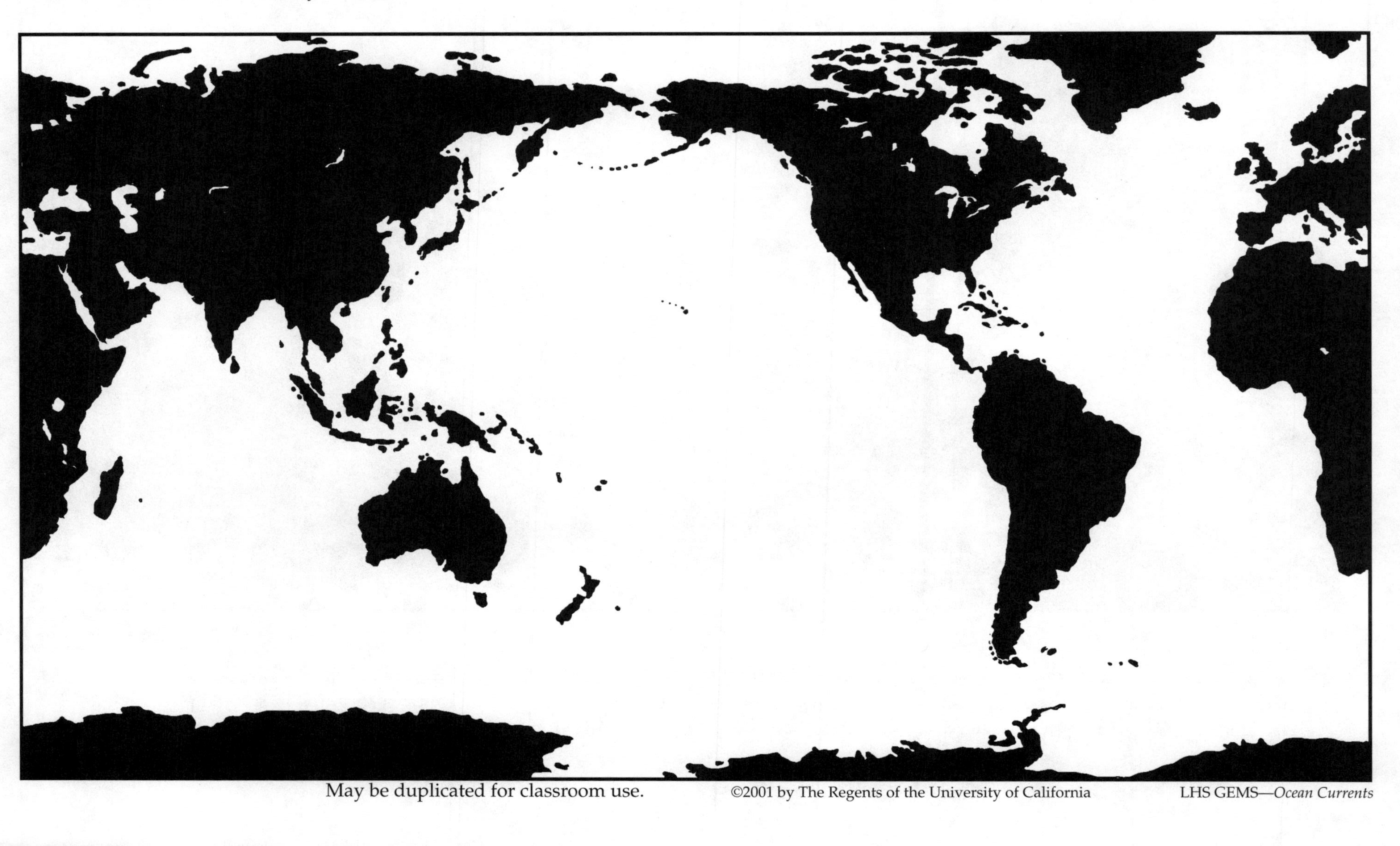

3. The Whaleship *Essex*

In 1820, the whaling ship *Essex* was rammed twice by a sperm whale. The ship sank, and three lifeboats were left adrift. Three months later, crew members from two of the lifeboats were rescued. The "X" on the map marks the spot where the ship sank. Mark with an **ORANGE** pen the route you think the lifeboats may have drifted, and where you think they were rescued.

Use **MAP 1** to mark your route.

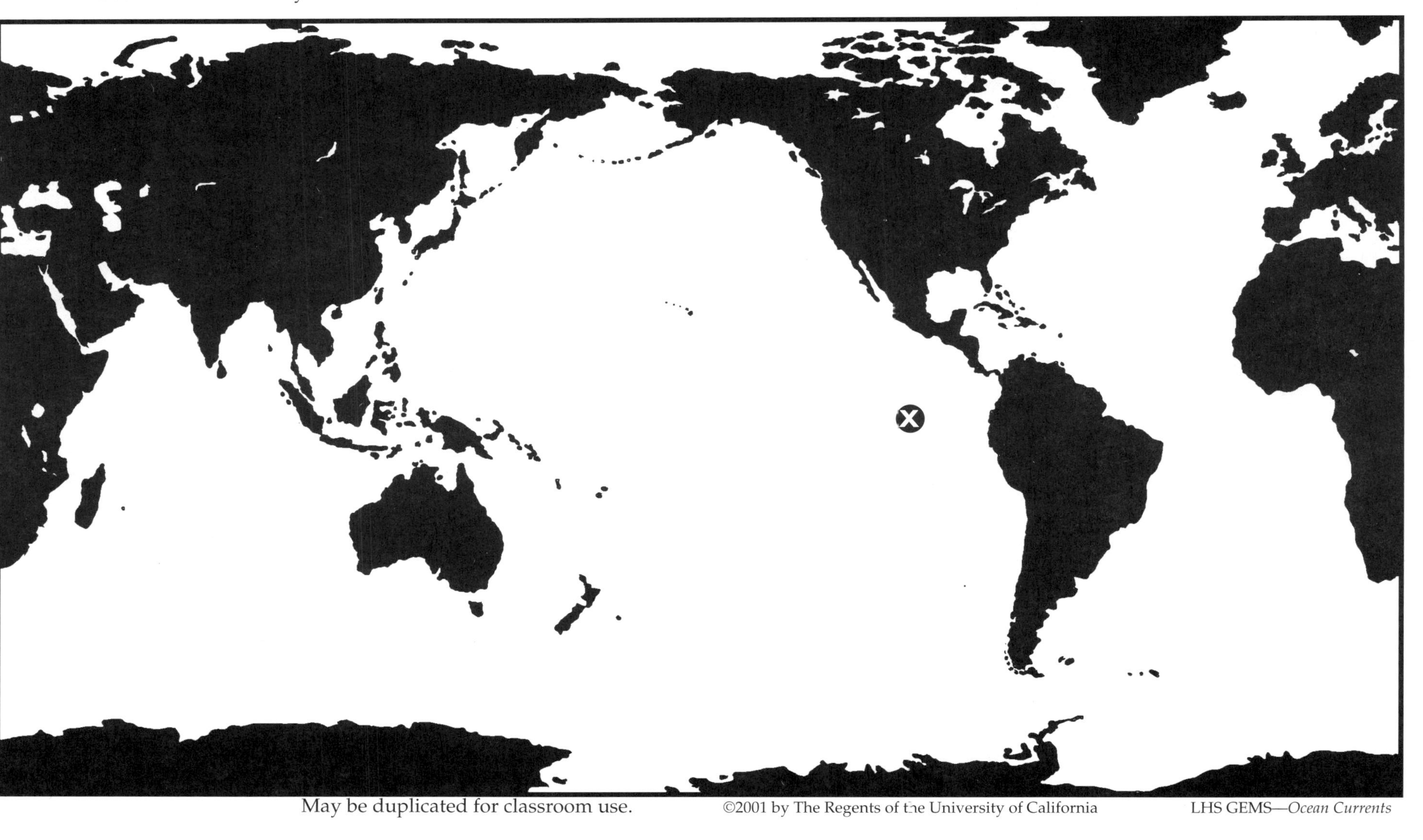

4. Nike Shoe Spill

In 1990, a ship traveling from Korea to Seattle lost five containers of Nike athletic shoes. More than 61,000 Nike shoes were left floating south of Alaska. The "X" on the map marks the spot where the shoes were lost. Mark with a **RED** pen the route along which you think these shoes might have drifted.

Use **MAP 2** to mark your route.

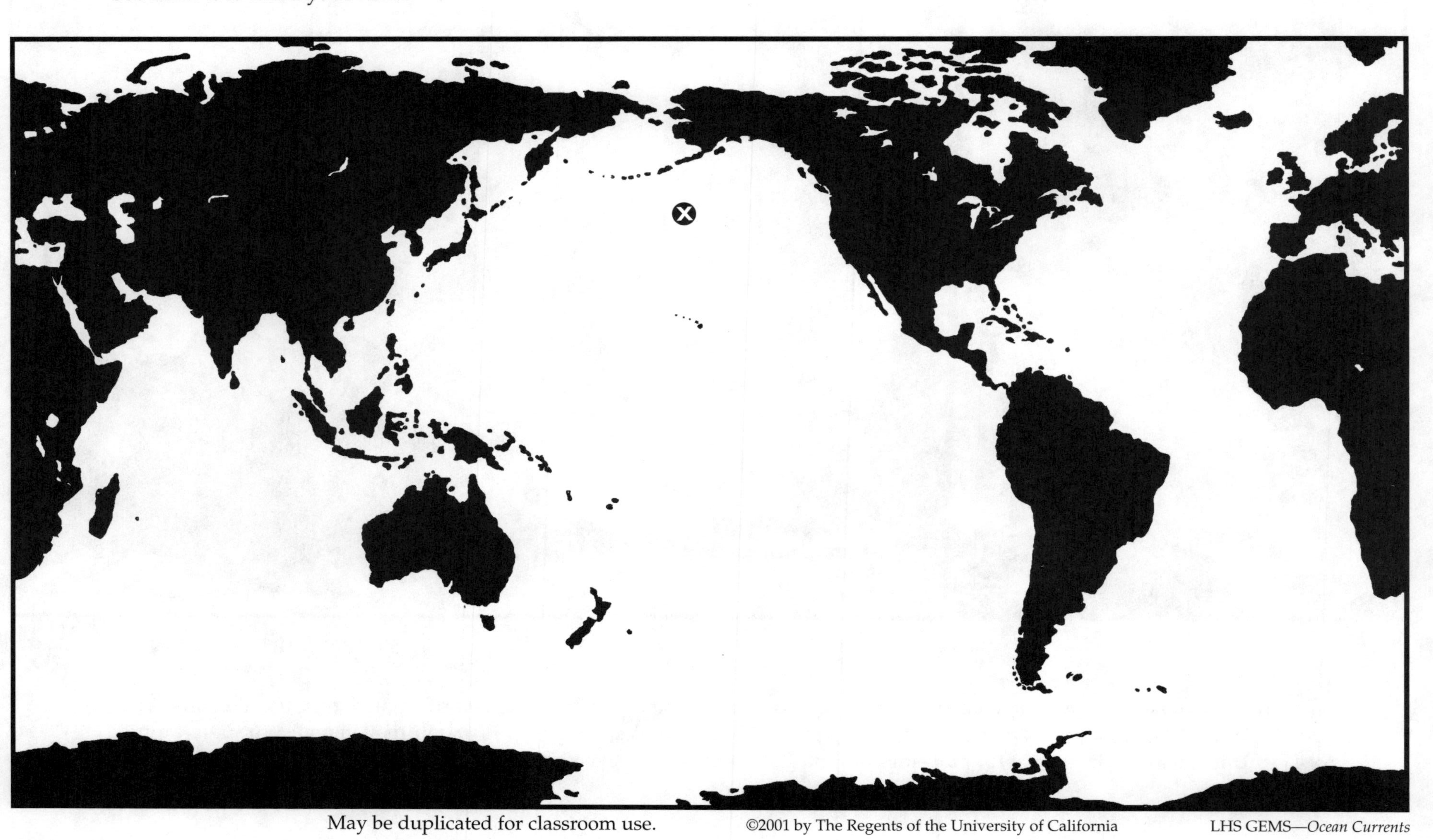

5. The *Endurance*

In 1914, Ernest Shackleton's ship got stuck in ice in the Weddell Sea in Antarctica, was slowly crushed, and then sank. He and his crew took three small boats and some food onto the ice. They drifted on the ice for five months, then dragged the boats to open water. They finally landed on an island at the tip of the Antarctic Peninsula—marked with an "X" on the map. There was no hope for rescue, and between them and any help was the most violent ocean area on Earth. Shackleton and five others set out for help in a 22-foot-long boat.

Mark with an **ORANGE** pen the route you think they should have taken to go for help. Use **MAP 2** to mark your route.

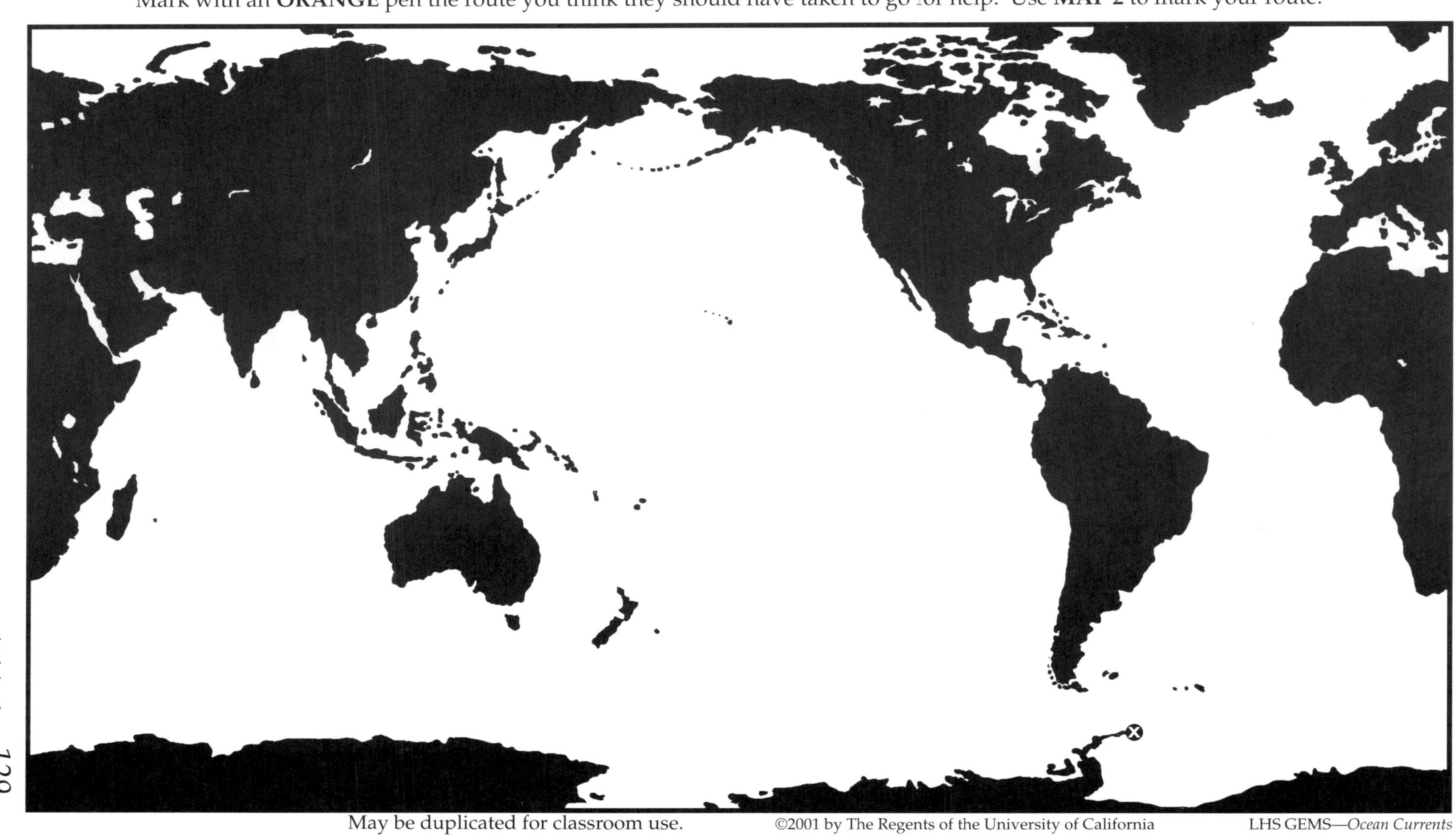

6. Swimming the Atlantic

Imagine you were going to swim across the North Atlantic Ocean, with a boat to sleep, rest, and eat on when you weren't swimming. Mark with a **BLUE** pen the route you think would be easiest.

Use **MAP 2** to mark your route.

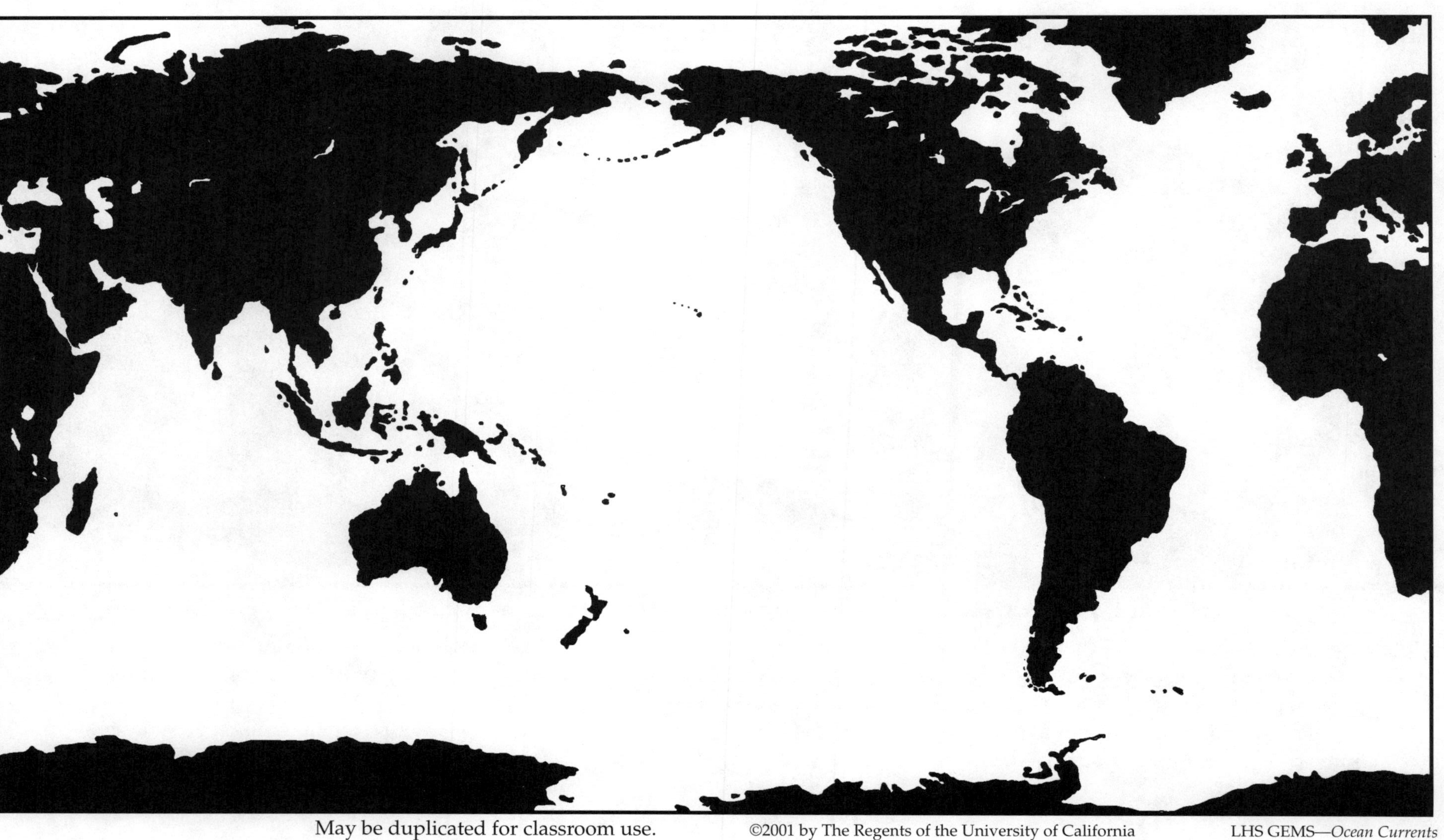

7. *Kon Tiki* and *Ra*

In 1947, 1969, and 1970, Norwegian researcher Thor Heyerdahl made voyages to test out his controversial ideas.

- People from North Africa settled in South America after sailing there in reed boats.
- People from South America settled in Polynesia, after sailing there on balsa rafts.

Use a **RED** pen to mark what you think is the easiest route from North Africa to South America and from South America to Polynesia. Use **MAP 3** to mark your routes.

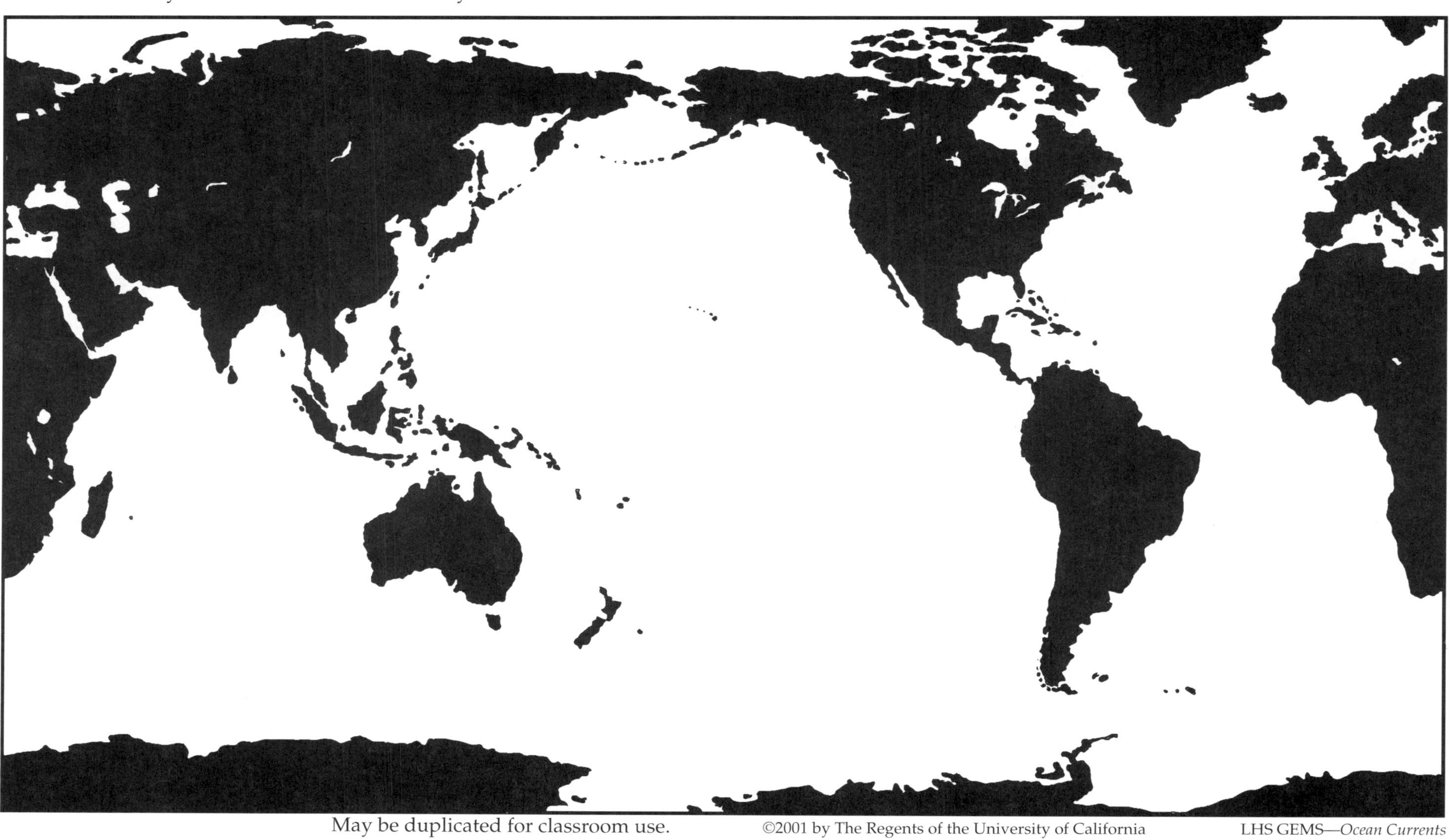

8. Rowing the Pacific

On July 11, 1991, 45-year-old Gerard d'Aboville rowed across the North Pacific Ocean.
Mark what you think would be the easiest route with a **BLUE** pen.

Use **MAP 3** to mark your route.

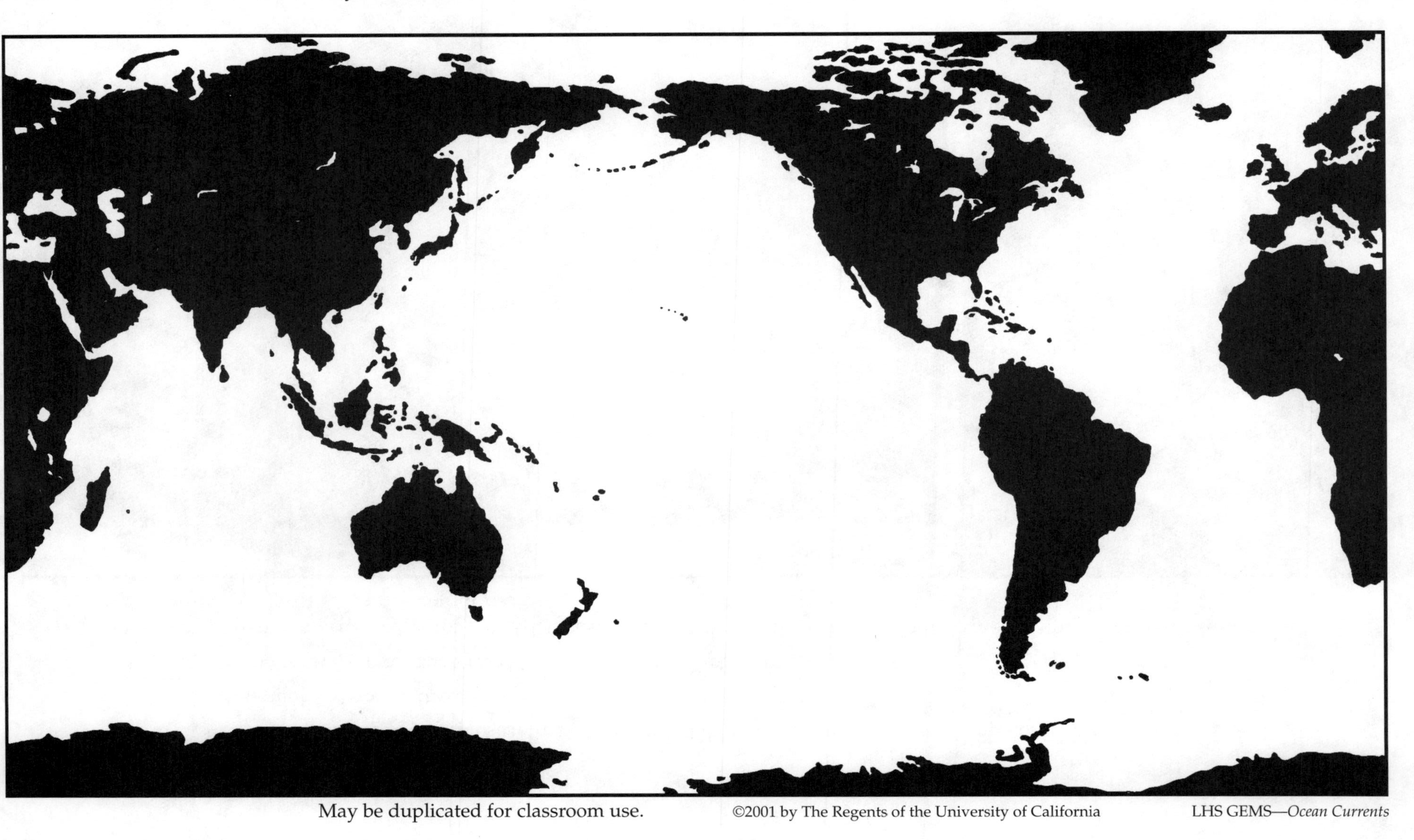

9. Penguin Feather

If a penguin feather with about the same density as water was carried underwater starting at the tip of the Antarctic Peninsula (marked with an "X" on the map), what route do you think it might take? Mark it with a **GREEN** pen.

Use **MAP 4** to mark your route.

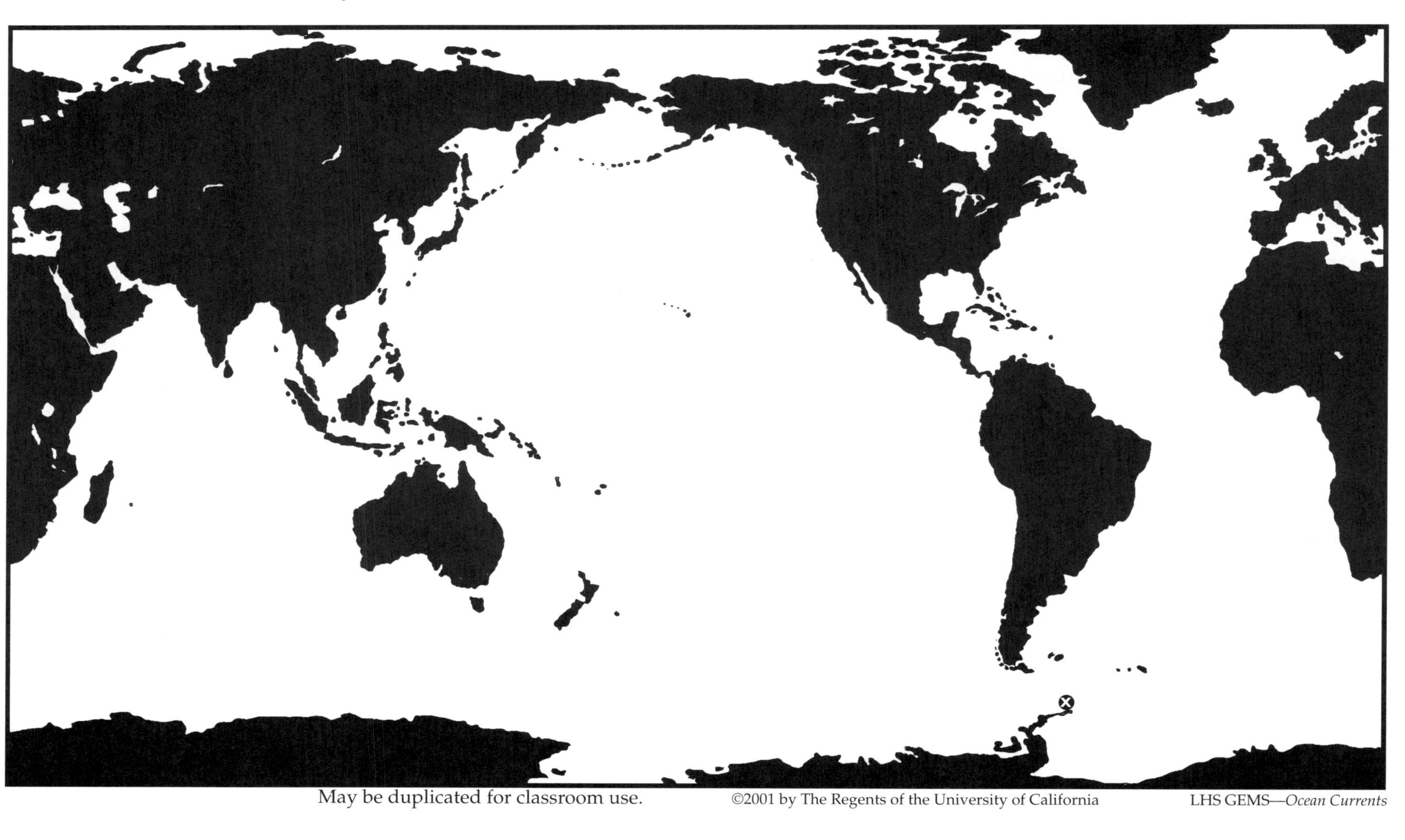

10. Rich Fisheries

Imagine you own a fishing boat. Mark on the map with a **RED** pen where you think the best fisheries in the world may be.

Use **MAP 4** to mark your fishing spots.

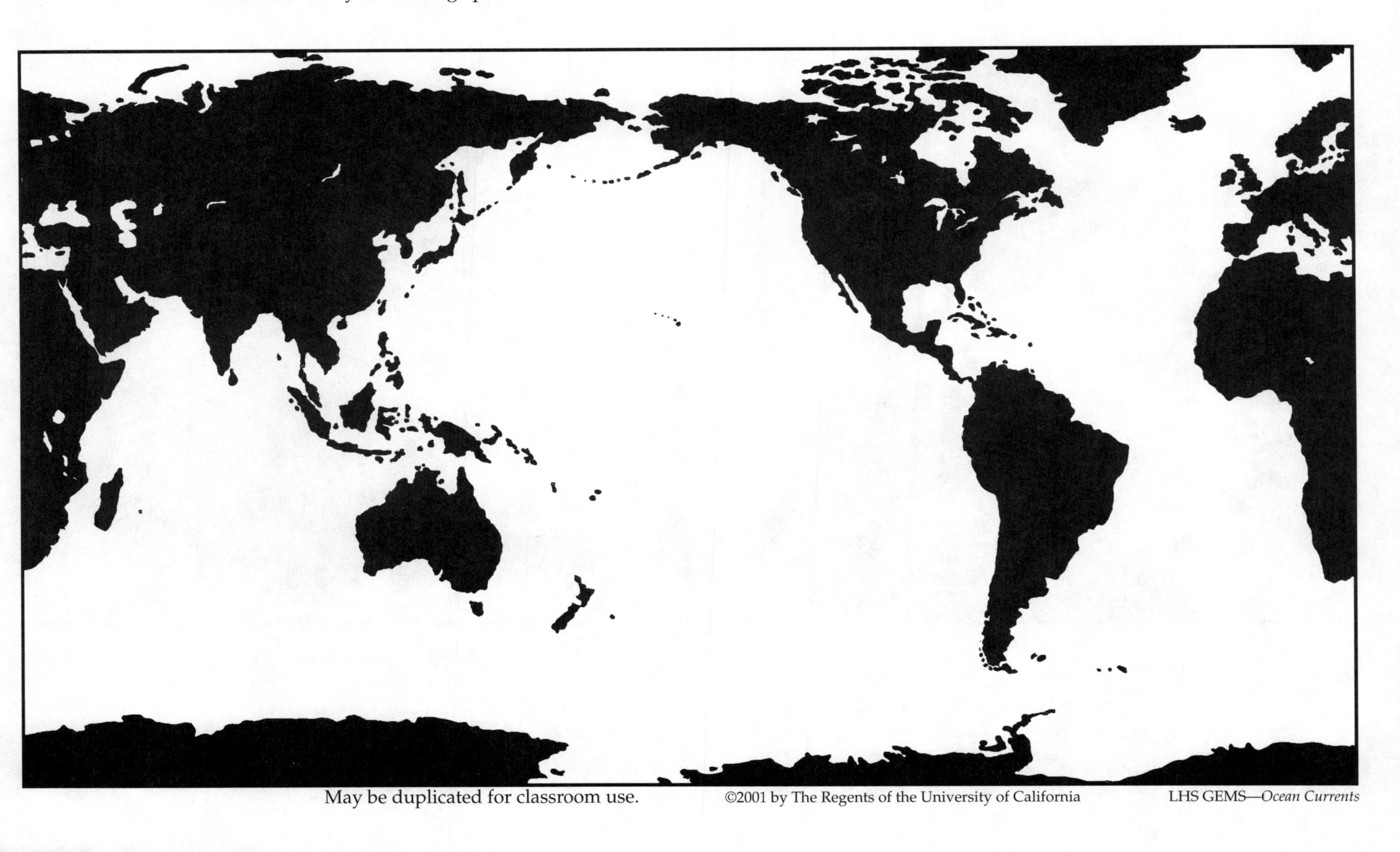

11. Waste Dumping

Imagine some hazardous wastes have been dumped in the two areas marked with an "X" on the map. Mark with a **BLUE** pen where you think they might move. Be sure to consider both deep (map 4) and surface currents (maps 1–3).

Use **MAP 4** to mark your route.

UPWELLING

DOWNWELLING

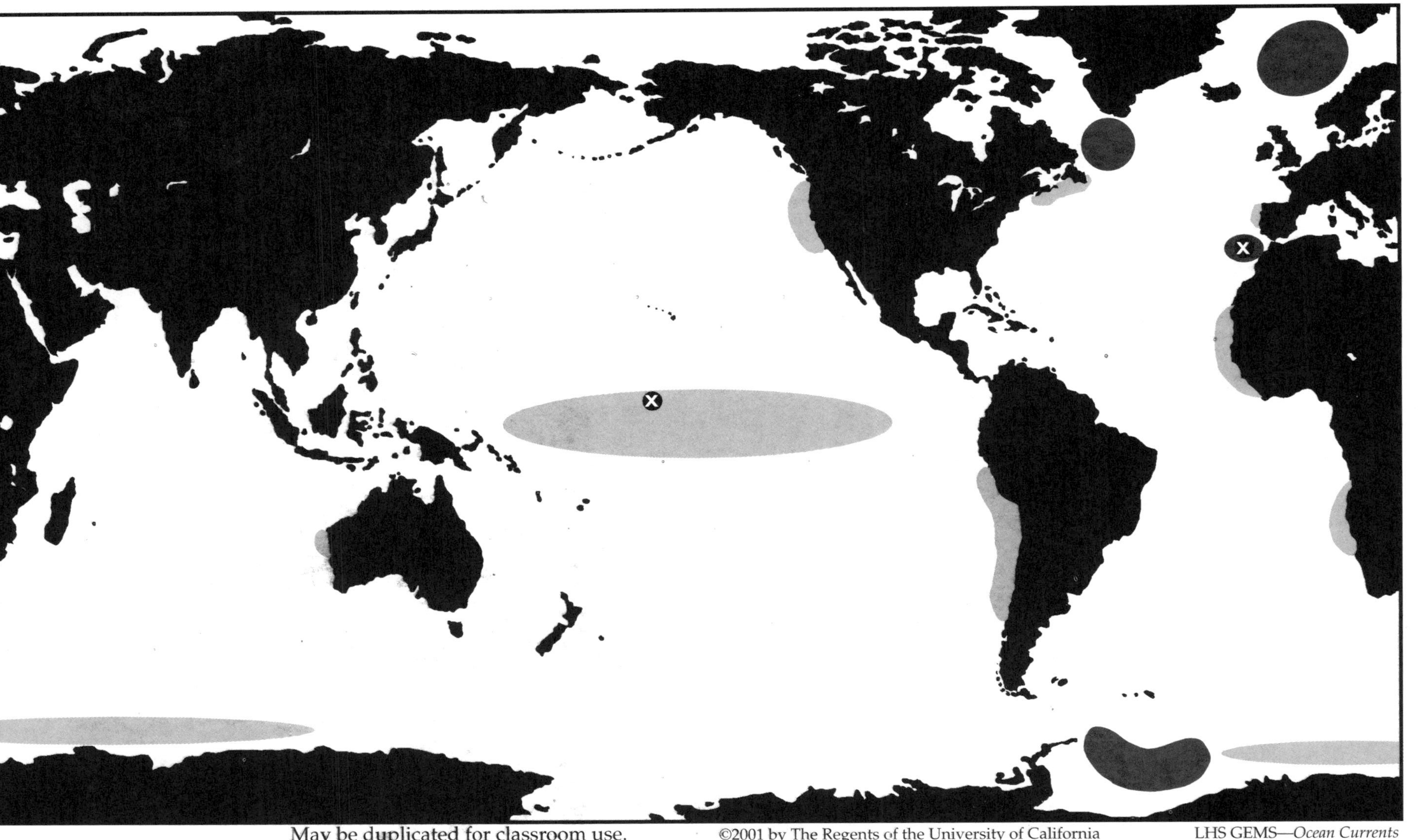

Once there was a man named Billy Bob. He was 22 and was on his way back to his hometown in 1999, March 17th.

Dear Journal,

It's me Billy Bob. I am going back to my hometown in California today. I am taking a big boat and a lot of people about 10-15. We are planning to start here in China and follow the currents and temperatures to San Diego, California. Well, I've got to go, hopefully there will be no big storms on the way.

Billy Bob

P.S. I will write when I get there.

Billy Bob went on his way and said good-bye to his dear friends. He was so excited about the trip. It was going to take him 10 days to get from China to San Diego California, but we'll see, Yes, we'll see.

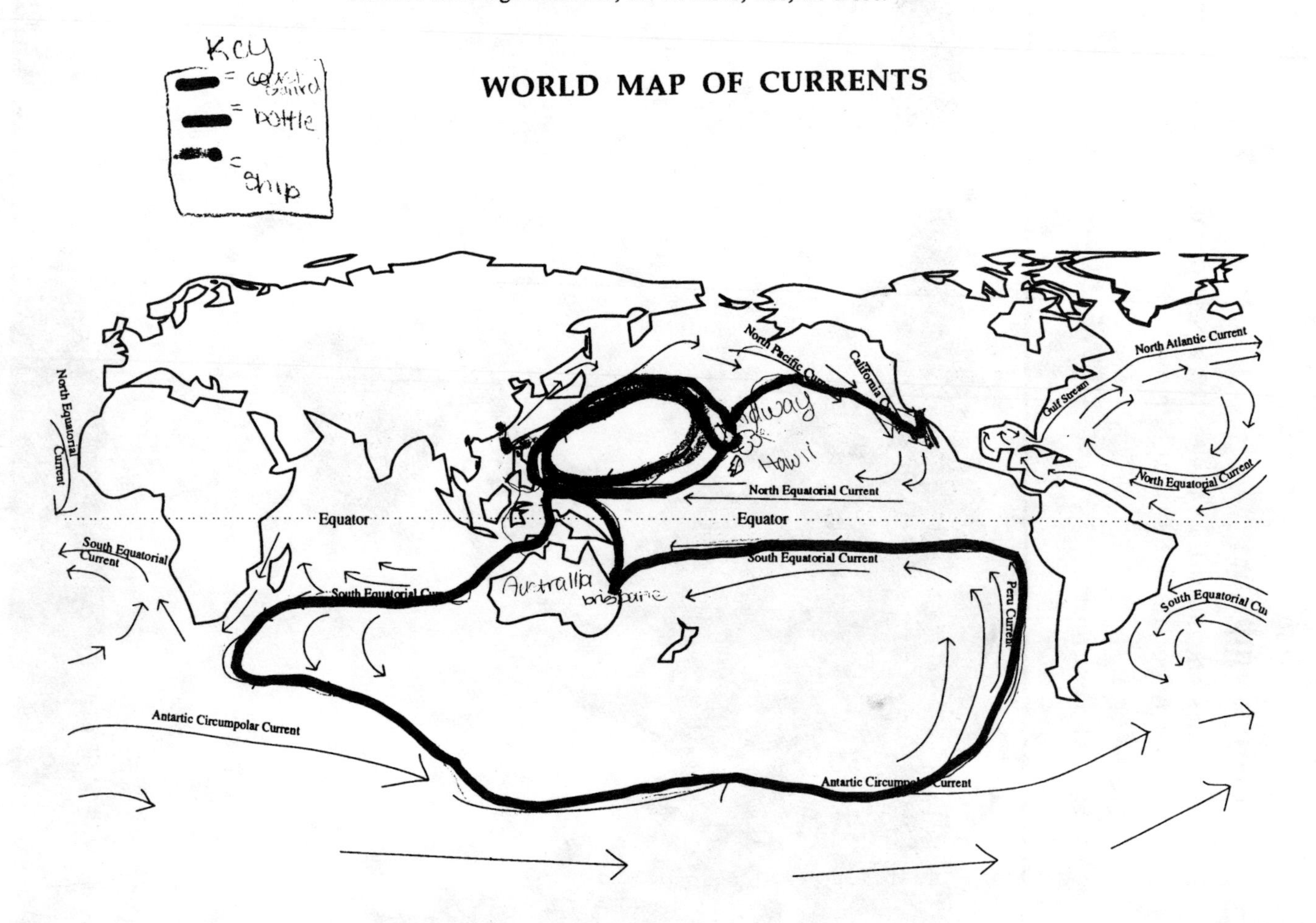

Activity 7: Message in a Bottle

Overview

Students use world currents maps and the knowledge they have gained to make up their own fictional stories involving ocean currents.

Message in a Bottle can serve as an excellent assessment activity for the *Ocean Currents* unit, as students are asked to use their stories to express the main things they have learned about the ocean. Specifically, they are asked to include information on ocean currents, their causes and effects; as well as discussing information relating to wind, density, temperature, and salinity.

The activity begins with students reviewing the previous activities through a gallery walk of the posters, charts, and Key Concepts generated earlier. They brainstorm a class list of the most important points and Key Concepts they think should be included in their creative story. This list serves as the class criteria on which their knowledge will be assessed.

What You Need

For the class:

- ❒ overhead transparencies of Surface Currents and Deep Currents maps from Activity 6, pages 109–110
- ❒ posters, charts, and Key Concepts from previous activities of the unit
- ❒ 1 sheet of chart paper
- ❒ marking pens, various colors

For each student:

- ❒ 3 or more sheets of lined 8½" x 11" paper
- ❒ pen or pencil
- ❒ colored markers
- ❒ 1 copy of either the Surface Currents or Deep Currents maps from Activity 6, pages 109–110

Getting Ready

1. Have current maps and writing materials on hand. Depending on the content of their story, students will need either the Surface or Deep Current map from Activity 6. If you plan to assign this activity as homework, students will need copies of both maps so they can choose which to use.

2. Post any charts, posters, and Key Concepts from the previous activities to use as graphic reminders of the important points and concepts covered earlier in the unit. If you or your students have gathered additional resource materials that may be helpful, have them available as well.

3. Decide if there are any specific aspects of the unit you want students to be sure to include or address in their stories. During the activity your ideas can be added to the list brainstormed by the students.

4. Determine if this will be a homework assignment or if the students will be given class time for writing.

Introducing the Activity

1. Remind students of the ocean current stories from the previous activity. Ask them if they've ever heard of any other interesting stories involving currents.

2. Let them know they will get to make up their own stories. First, the class will help determine what points and concepts would be important to include in a story if it were to be used to find out what they have learned about ocean currents.

3. Point out the posters, charts, and Key Concepts displayed around the room. Have the class work in small groups and travel around the room in a "gallery walk" looking at the items posted on the walls. Have one person in each group act as the recorder and make a list of what they feel are the 10 most important ideas, concepts, or points they think would be important to include in a story about currents.

4. After each small group has made their list, lead a class brainstorm and record each of their ideas on chart paper. If students repeat a point already detailed on the chart, put a star by it to show that more than one group thought it was important to include on the class list.

5. Have the class help you highlight those points the entire class thinks are important to include in their stories. At this juncture you might want to add some additional points if you feel some important concepts were overlooked.

6. Tell the class the highlighted points are those you will be looking for in their creative stories.

Clarifying the Task

1. Explain that their stories can be about shipwrecks, people lost at sea, people from ancient times, animals washed to sea on a raft of tangled trees, or difficult sea journeys. They could also be about travels of inanimate objects, such as a message in a bottle, spilled cargo of athletic shoes, a pollutant spill, or similar subjects. The assignment could be written as a story or an essay, a journal or log, a legend, or a firsthand account.

2. Tell students their story must include **accurate information about ocean currents** as detailed on the class list, and, while their stories will be evaluated primarily on this information, creativity counts too. They should draw upon what they learned in the unit—not only the recent stories about voyages, but also the stations concerning density, temperature, and salinity, the wind demonstrations, the waste disposal model, and any other activities or information.

3. Project the transparencies of the current maps students will use. Let them know they need to draw the route from their story on the map, and attach it to their story. They can use either the Deep Currents or Surface Currents map, depending on the content of their story.

4. Give them an example or two to get them started in their writing, such as the following:

- They could write a story about a message they placed in a bottle and set loose at some location of their choice. They would provide a map showing the route and a written explanation detailing how, if they set the bottle loose at one location, ocean currents they specify would enable it to go to an intended distant location somewhere else on the globe.

- Have them pick any spot in the ocean and follow a likely route from that spot. They can create a story of who or what traveled that route and what sort of adventures were encountered along the way. Again they would need to include a map showing the route and a written explanation including the important concepts listed by the class.

Ask students what language they should write their message in so it can be understood by the people most likely to find it.

Creating the Stories

1. Distribute the maps, paper, and writing utensils and let the students begin.

2. Circulate as needed to answer questions and remind students that their stories should include as much information about ocean currents as possible.

3. Determine how you want students to share their stories with each other and the class. These stories could be used to show parents what students have learned, or as displays during an Open House at school.

Behind the Scenes

What Are Ocean Currents?

Ocean currents are usually defined as the horizontal and vertical circulation system of ocean waters produced by gravity, wind friction, and water density variation in different parts of the ocean.

Currents shown on maps are highly simplified and show average conditions. But real currents are like real winds; they vary a lot from the average.

The direction and form of oceanic currents is governed by a number of natural forces:

- Coriolis effect caused by the rotating Earth
- Friction caused by winds blowing over the ocean's surface as well as the friction between different layers of water
- Variable density of seawater, which results mainly from temperature and salinity variation

The Coriolis effect causes the major ocean currents to form clockwise circulations in the Northern Hemisphere and counterclockwise circulations in the Southern Hemisphere. These circulatory systems are called **gyres**.

In addition to horizontal current movement, **vertical oceanic circulation** is important because it brings up deep ocean waters and moves surface water down. Partly caused by the effects of wind near the surface, vertical circulation also occurs because of variations in salinity and temperature. Saltier water is more dense than fresher water, and warm water is less dense than cold water. Water that is more dense tends to sink, and less dense water tends to rise. So density differences can cause vertical currents.

Currents and Climate

The study of currents revealed that ocean currents profoundly affect weather and climate. For example, the Gulf Stream-North Atlantic-Norway Currents bring warm tropical waters northward and eastward, warming the climates of eastern North America, the British Isles, and Ireland, as well as the Atlantic Coast of Norway in winter. The Kuroshio Current and the North Pacific Current do the same for Japan and western North America, where warmer winter climates also result. The warmer waters also evaporate more readily in the warmer temperatures and help generate increasing rain and snow fall along these coasts.

As oceanic voyages became common in the 18th century, the study of oceanic currents intensified in order to improve sailing ship routes.

On the other hand, the cold Peru, California, and Benguela Currents hinder evaporation and, as they flow along the warmer coasts of Southern California, South America, and southwest Africa, they create fogs but no rain, resulting in the dry deserts of California, Peru, Chile, and Namibia. But these cold currents are fed in part by cold waters welling up from below. These waters are rich in nutrients, and some of the world's best fishing grounds are found in them. Ocean currents and atmospheric circulation influence each other. For example, periodically the trade winds above the tropical Pacific weaken and the eastern tropical Pacific waters are warmed, creating an El Niño event, which widely affects climate and weather, bringing drought to Australia, more winter storms to Southern California, and a warm winter to northern North America.

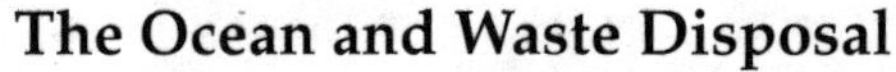

The Ocean and Waste Disposal

One of the least known, but most significant, uses of the ocean is as an enormous dump site. In the past, the ocean seemed able to assimilate the wastes of society. Industrialization and related developments, as well as increases in global population, have given rise to huge amounts and types of waste that are causing major ocean pollution. Some areas of the ocean are heavily polluted by human wastes—from the raw sewage of cities to junked appliances and cars. Less apparent but more insidious forms of pollution are toxic chemicals, nuclear wastes, and oily bilge water pumped overboard into the ocean by practically all vessels using petroleum for power.

Massive oil spills from tanker accidents have severely damaged beaches and estuaries and caused widespread harm to wildlife. Large power plants are often located along coastlines and negative environmental effects can be caused in the immediate area of the power-plant outfall, such as disposing of the plant's cooling water. Fertilizers, animal wastes, herbicides, and pesticides reach the oceans through the wind and rivers, and cause excess algal growth, oxygen depletion in the bottom waters, and contamination of marine organisms. The edges of the ocean—beaches, lagoons, and bays—are considered the most sensitive to human impact, but the continued dumping of wastes can eventually impact much larger areas of the ocean.

The Coriolis Effect and Upwelling

The Coriolis effect represents the effects of the rotation of the Earth. It is one of the factors involved in creating the major ocean current gyres that spin clockwise in the Northern Hemisphere and counterclockwise in the Southern Hemisphere. The Coriolis effect is also important in causing coastal upwelling.

The Coriolis effect was first described by the French mathematician and physicist Gaspard Coriolis in the early 19th century. He recognized that an object that moves in a straight line above the surface of the Earth will appear to people on the rotating Earth to curve because the Earth is turning under it. Movement in the Northern Hemisphere is diverted to the right (if we look in the direction the object is moving) and to the left in the Southern Hemisphere.

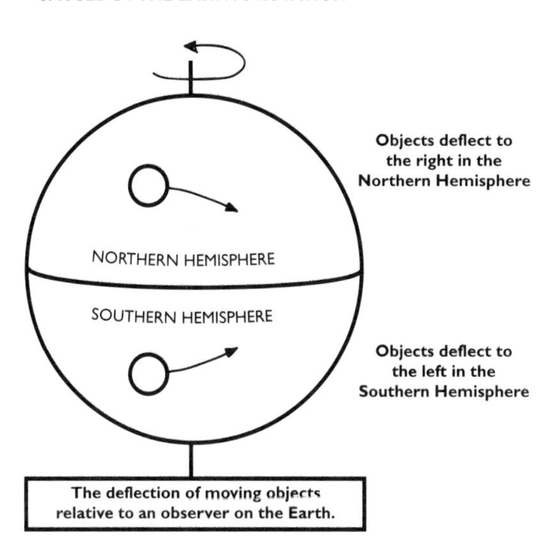

Demonstrating the Coriolis Effect and Upwelling

You can demonstrate this effect and its influence on currents and the resultant upwelling. Use the large inflatable globe (or another globe) and tell the class that you will try to draw on the globe the direction you would assume water would be set in motion based on the direction of the wind. For example, one would assume that wind blowing from the north along the California coast would set the water in motion down the coast. Attempt to draw this current on the globe, but as you draw also turn the globe towards the east to represent the spinning of Earth.

There are a number of resources on the Internet that explain and demonstrate the Coriolis effect—sometimes known as the Coriolis force—as relates to winds, weather, and currents. For example, www.windpower.dk.

Show the students the line that results from your drawing. They will notice that the line actually ended up going towards the west instead of down the coastline as the Earth "turned out from under the water." The net effect is that surface water is moved offshore **allowing water from the depths to rise up (upwell) to the surface.**

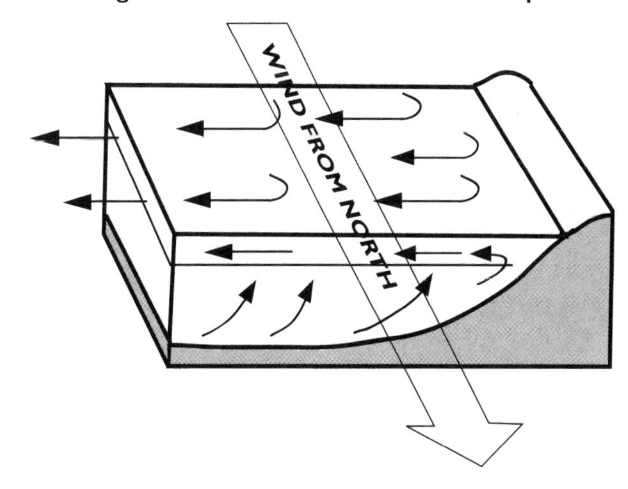

Point out on the map or globe the following major areas where coastal upwelling is found. The west coasts of North America and South America, North Africa and South Africa, and Australia. It also occurs off Spain and Portugal and along the coast of Antarctica.

Point out the Galapagos Islands off Ecuador, and draw their attention to how close it is to the equator. Let them know that although it would seem that the water should be quite warm there, it is only about 20°C at the surface. Challenge them to try to explain why. In general there is upwelling along the equator as surface water is deflected to the right in the Northern Hemisphere and to the left in the Southern Hemisphere to be replaced by water from below.

This background section is only a general introduction to the causes of upwelling. Please see "Resources" on page 145 for more complete information.

EQUATORIAL UPWELLING

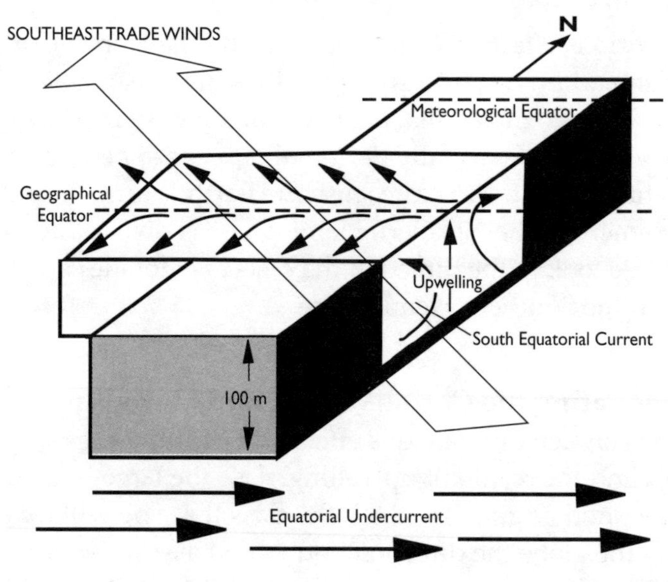

Cold Water and Nutrients

Cold water tends to hold more nutrients (e.g., nitrogen, potassium, phosphorus) and so is generally much more productive than warm water if it occurs in the sunlit upper layer of the ocean—the photic zone. Along much of the west coast of North America, for example, the winds often blow from the northwest and help push the cold, rich surface waters of the California Current slowly down the coast. All year long, animals and plantlike organisms die and decompose and slowly sink toward the bottom out of the sunlit photic zone—this is sometimes called the "rain of bodies." In the spring and summer (March through September), the winds from the northwest are at their strongest along much of the west coast of North America. This is when upwelling is strongest in these regions.

Blue whales are the largest animal ever to live on our planet. They eat up to two tons per day of krill or small fish. One of the world's largest remaining blue whale populations chooses the California coast as its feeding grounds each summer.

Upwelling caused by winds off the California coast in spring and early summer, chills the air, contributes to San Francisco's fog banks and cool summers, and makes the waters there some of the most productive in the world. The central California coast, because of upwelling, is one of the most productive places on Earth.

Resources

Materials

Activity 1: Globes showing ocean currents

AMAC Plastics
740 Southpoint Blvd.
Petaluma, CA 94954-3832
(707) 763-3366

Activity 1: Marine Mammals of the Gulf of the Farallones poster

Gulf of the Farallones National Marine Sanctuary
Fort Mason, Building 201
San Francisco, CA 94123
(415) 556-3509

Activity 1: East Coast poster

Jacques Cousteau National Estuarine Research Reserve
Institute of Marine and Coastal Sciences (IMCS)
Rutgers University
71 Dudley Rd.
New Brunswick, NJ 08901

Activity 1: Gulf Coast (Barrier Islands) poster

Carolina Biological Supply Company
2700 York Rd.
Burlington, NC 27215
(800) 334-5551
item no. BA-57-4727

Activity 2: Colored images of ocean temperatures and currents

SeaWiFS Project Image Archive
http://seawifs.gsfc.nasa.gov/SEAWIFS/IMAGES/GALLERY.html

Ocean Currents
http://www.athena.ivv.nasa.gov/curric/oceans/drifters/ocecur.html

General Characteristics of the World's Oceans: Ocean Currents
http://icp.giss.nasa.gov/research/oceans/oceanchars/currents.html

Activity 2: Dissolved Oxygen Test Kit

Carolina Biological Supply Company
2700 York Rd.
Burlington, NC 27215

(800) 334-5551

item no. BA-65-2865

LaMotte Chemical Company
P.O. Box 329
Chestertown, MD 21620

(800) 344-3100

Activity 3: Tornado Tube

Also available from science museums, science stores, novelty stores, and some scientific supply companies.

Tornado Tube
26 Dearborn St.
Salem, MA 01970

(508) 745-1788

Activity 3: Hydrometer

Make your own using the directions on page 62, or order from

Carolina Biological Supply Company
2700 York Rd.
Burlington, NC 27215

(800) 334-5551

item no. BA-67-1935

Rheoscopic fluid, a nontoxic and reusable substance for flow visualization, can produce striking visual images of currents within a liquid. Rheoscopic fluid is available from its maker, Novostar Innovations for Education, in Burlington, North Carolina, at (800) 659-3197, as well as from other distributors, including the authorized GEMS Kit™ distributor, Sargent-Welch. The National Science Teachers Association (NSTA) publishes an activity guide entitled Project Earth Science: Physical Oceanography, *by Brent A. Ford and P. Sean Smith, which outlines several class activities about ocean currents using this intriguing and revealing fluid.*

Related Curriculum Material

Living in Water: An Aquatic Science Curriculum for Grades 5–7, National Aquarium in Baltimore, Kendall/Hunt Publishing, Dubuque, Iowa, 1997

Polynesian Voyaging & The Wayfinding Art: A Comprehensive Curriculum of Activities and Information for Teachers and Students, Gail K. Evenari, 1995

Project Earth Science: Physical Oceanography, Brent A. Ford and P. Sean Smith, National Science Teachers Association, Arlington, Virginia, 1995

Books

Drift Bottles in History and Folklore, Dorothy B. Francis, Ballyhoo Books, Shoreham, New York, 1990

The Gulf Stream: Encounters with the Blue God, William MacLeish, Houghton Mifflin, Boston, 1989

Through a very nontechnical approach, MacLeish tells the story of the Gulf Stream and its importance.

Ice Story: Shackleton's Lost Expedition, Elizabeth Cody Kimmel, Clarion Books, New York, 1999

Describes the incredible events of the 1914 Shackleton Antarctic expedition, when the ship *Endurance* was crushed in a frozen sea and the men made the perilous journey across ice and stormy seas to reach inhabited land. Contains many photos taken by the ship's photographer.

In the Heart of the Sea: The Tragedy of the Whaleship Essex, Nathaniel Philbrick, Viking Penguin, New York, 2000

Recounts the story of the wreck of the whaleship *Essex* and of its crew's struggle for survival. Woven into the story is the history of Nantucket (home port of the ship) and the dangers of the whaling industry. The disaster was the inspiration for the climax of *Moby Dick.*

I Wonder Why the Sea is Salty and Other Questions About the Ocean, Anita Ganeri, Kingfisher, New York, 1995

Kon-Tiki: Across the Pacific by Raft, Thor Heyerdahl, Amereon Limited, Mattituck, New York, 1993

To prove his theories on the settlement of the South Sea islands, Heyerdahl and five others sailed from Peru on a balsa log raft. After three months at sea—facing storms, whales, and sharks—they arrived at a Polynesian island.

The Mysterious Ocean Highway: Benjamin Franklin and the Gulf Stream, Deborah Heiligman, Raintree Steck-Vaughn, Austin, Texas, 1999

Describes the study of the Gulf Stream from Benjamin Franklin's days to our own and explains its characteristics.

Oceans, Seymour Simon, Morrow Junior Books, New York, 1990

Discusses the Earth as a planet covered with water on which the continents are islands. The topography of the ocean floor is discussed as are wave motion, wave energy, and ocean currents.

Oceans for Every Kid, Janice VanCleave, John Wiley & Sons, New York, 1996

This overview of marine sciences includes information on techniques and technologies of oceanography, the topology of the ocean floor, movement of the sea, properties of sea water, and life in the sea.

Oceans: Looking at Beaches and Coral Reefs, Tides and Currents, Sea Mammals and Fish, Seaweeds and Other Ocean Wonders, Adrienne Mason, Kids Can Press, Buffalo, New York, 1995

Detailed drawings and full-color photos bring the ocean to the reader while safe and simple experiments and activities help the reader explore the mysteries of the Earth's oceans and ocean life.

Our Oceans: Experiments and Activities in Marine Science, Paul Fleisher, Millbrook Press, Brookfield, Connecticut, 1995

An introduction to the study of marine science with separate chapters on seawater physics and chemistry, geology (including recent discoveries about the ocean floor), ocean currents and their effect on world weather, and undersea resources and the need to conserve them.

Polluting the Sea, Tony Hare, Franklin Watts, New York, 1991

An easy to understand introduction to a large topic, this discusses various forms of pollution—oil, litter, sewage, metals, chemicals, and radioactivity—and explains why it happens, what effects follow, and what can be done about the problem.

The Restless Sea, Robert Kunzig, W.W. Norton, New York, 1999

Through compelling and spellbinding writing, this book reports on what is known about the world's oceans and how that knowledge was assembled over the centuries. Especially relevant to *Ocean Currents* are Chapters 9 and 10 in which such topics as deep currents, surface currents, gyres, eddies, convection currents, and thermohaline circulation and its effect on global climate are discussed.

The Seas in Motion: Waves, Tides and Currents—How They Work; Their Causes and Effects, F.G. Walton Smith, Thomas Y. Crowell, New York, 1973

Waves, Tides and Currents, Daniel Rogers, Bookwrights Press, Charlottesville, Virginia, 1991

An introduction to waves and erosion, tsunamis, tides, floods, and currents—how they are created and their effect on the climate and the shorelines. Illustrated with well-chosen color photos and diagrams.

Magazine

Ocean Explorer, Vol. 3, No. 4, May 1994, Woods Hole Oceanographic Institution

The theme of the entire eight-page issue is "The Ocean in Motion." For information on issues, write to Ocean Explorer, Associates Office, Woods Hole Oceanographic Institution, Woods Hole, MA 02543.

"Message in a Bottle," Kevin Krajick, *Smithsonian*, Vol. 32, No. 4, pages 36–47, July 2001

Reseacher Curtis Ebbesmeyer traces the flow of ocean currents the old-fashioned way—by studying the movements of random junk, such as lightbulbs, toilet seats, and rubber duckies, cast up on shores around the globe.

Videos

Ocean Drifters, National Geographic Video/Columbia TriStar Home Video, Culver City, California, 60 min., 1993

This documentary looks at the many species of marine life that use the ocean currents to travel—nearly effortlessly—around the globe.

Understanding Oceans, Discovery Channel, 51 min., 1997

Most of the water on the planet is gathered in large, salty oceans that are pushed by the wind, pulled by the moon, and swirled as the planet spins. Explore this hidden world and meet some of its endangered creatures.

Internet

Around the World
www.naau.com
www.goals.com/

Features people who are sailing, kayaking, and diving around the world!

The Bridge: A Web-based Resource for Ocean Sciences Educators
www.vims.edu/bridge/

A clearinghouse of the best K–12 ocean sciences education sites available online.

Cousteau Society
www.cousteausociety.org

Information about the society that has helped millions of people understand and appreciate the fragility of life on our water planet through its explorations and observations of ecosystems throughout the world. Includes Dolphin Log in the Classroom educational program.

Global Drifter Center at Atlantic Oceanographic and Meteorological Laboratory
www.aoml.noaa.gov/phod/dac/gdc.html
Shows actual data and maps of drifter buoys.

InciteScience!
www.incitescience.com
A science education Web site, including a marine sciences literature and resource list.

Investigating Currents
virga.sfsu.edu/courses/geol102/ex8.html

A Map of Major Oceanic Surface Currents
www.acl.lanl.gov/GrandChal/GCM/currents.html

MARE
www.lhs.berkeley.edu:80/MARE/
The Marine Activities, Resources & Education (MARE) program at the Lawrence Hall of Science is a dynamic, K–8, multicultural science program that transforms entire elementary and middle schools into laboratories for the exploration of the ocean. This schoolwide interdisciplinary program explores different marine environments around the world from the perspective of the diverse cultures living around them. For more information call (510) 642-5008.

Mick Bird
www.goals.com/transrow/index.html
Mick Bird wants to be the first man to row around the world solo! In his rowboat *REACH,* Mick has challenged ocean winds and currents to row from California to Hawaii, Hawaii to the Marshall Islands, Marshall Islands to Australia/Indonesia. In June 2001 he hopes to launch the fourth leg of his Trans-Oceanic Rowing Expedition: Indonesia to Africa.

NASA Educational Site
www.athena.ivv.nasa.gov/curric/oceans/drifters/index.html
Contains information about tracking drifter buoys, activities for plotting drifter data, maps, and information on currents.

NASA Ocean Science Research
topex-www.jpl.nasa.gov/

National Climate Data Center & National Oceanic and Atmospheric Administration
www.ncdc.noaa.gov/ol/climate/globalwarming.html

National Marine Fisheries Service of National Oceanic and Atmospheric Administration
www.nmfs.noaa.gov

Ocean Currents
seawifs.gsfc.nasa.gov/OCEAN_PLANET/HTML/oceanography_currents_1.html
Part of the Smithsonian's Ocean Planet Online Exhibit (http://seawifs.gsfc.nasa.gov/ocean_planet.html) which contains a full exploration of the ocean environment.

Ocean Mixing
earthobservatory.nasa.gov/Study/LovelyDarkDeep

Tori Murden
www.oceanrowing.com/first_solo/tori/tori's_progress.htm
In December 1999, a 36-year-old attorney from Kentucky, Tori Murden, became the first woman—and the first person from the United States—to row across the Atlantic Ocean. This was her second attempt; during her first effort, in 1998, she narrowly escaped death when her boat was capsized 11 times by a hurricane. She was rescued by a merchant ship. She narrowly missed another hurricane during her second, successful attempt, though it did slow her down. The trip took her 81 days, 7 hours, and 31 minutes, from the Canary Islands just off the coast of Africa to the French-Caribbean island Guadeloupe. Your students may want to learn more about Tori Murden's achievement and about how the ocean currents influenced her route.

Other

Drift Bottles

Since 1986, 5th and 6th grade students at Harbor Day School in Corona del Mar, California, have launched about 200 message-filled bottles. The project was begun by science teacher Judy d'Albert who was inspired by an article she saw in *National Geographic World.*

The bottles her students use—called drift bottles or drogues—are glass, sealed with cork and wax. Before launching their drogues, d'Albert's students discuss such topics as geography, currents, prevailing winds, famous voyages, and environmental concerns.

Most of the drogues are launched into the Pacific far off the coast of Southern California. Bottles have traveled to Mexico, Hawaii, and the Philippines (most have landed within 100 miles of each other). They float for thousands of miles and take 2–3 years.

One student arranged for bottles to be launched off the Olympic Peninsula in Washington. Another student launched several bottles from Fiji. In 1990, d'Albert dropped a bottle in the Atlantic from the West Indies. It floated for nine months and landed in Cuba. It was found by a man who, years later, emigrated to the United States with the help of his new friends at Harbor Day School.

If you would like more information about Judy d'Albert's project, you may contact her at jdalbert@harborday.org

Beachcombing

Dr. Curtis Ebbesmeyer, an oceanographer from Seattle, publishes the newsletter *Beachcombers' Alert.* In it, beachcombers around the world exchange information about what is floating in the ocean and what is landing on the beaches. Write to Dr. Ebbesmeyer, 6306 21st Ave. NE, Seattle, WA 98115 or visit his Web site at www.beachcombers.org

Assessment Suggestions

Selected Student Outcomes

1. Students demonstrate an increased awareness that "there is only one ocean" and that Earth's surface is covered by an interconnected ocean that circulates around all the continents.

2. Students improve in their understanding of ocean currents and their impact on human settlement, migration, trade, ocean life, and world climate.

3. Students are able to discuss and describe the main causes of ocean currents, especially wind and the interaction of water masses of different temperatures and salinities.

4. Students increase their practical understanding of the important scientific concept of density and are able to relate differences in salinity and temperature to density and to the formation of currents.

5. Students gain in their awareness of issues related to waste disposal and pollution of the ocean, and are able to describe how currents can spread pollutants throughout the ocean.

Built-In Assessments

Global Exploration. The global exploration worksheet students use in Activity 1 can serve as an indication of basic student understanding of the "only one ocean" concept. It also provides a great deal of other information on student knowledge. (Outcome 1)

Currents and Waste Disposal. As students predict and make tests with the waste disposal model and present findings as to the best and worst disposal locations, the teacher can see how well they've understood how currents spread pollution throughout the ocean. (Outcome 5)

Station to Station Observations and Poster. Teachers can observe how students approach the challenges and communicate as they rotate through the stations in Activity 3. More information on how well students understand the relation of temperature, salinity, and density to currents can be gained when student groups present their poster on a specific station. (Outcomes 3 and 4)

Delving Into Density. Teacher observations of students layering liquids in Activity 4 and of their participation in the follow-up discussion can reveal levels of student understanding of density. (Outcome 4)

Demonstrating Ice Cubes. As students predict whether an ice cube will melt faster in fresh water or salt water, then observe and hypothesize about the results, teachers can informally observe how well they are able to apply what they've learned to a new situation. (Outcome 4)

Think, Pair, Share. This student discussion format, which helps students sum up results and conclusions of the unit so far, can provide teachers with a sense of how much students have learned and how well they are able to articulate the main lessons and Key Concepts in their own words. (Outcomes 1, 3, 4, 5)

Charting Real Routes. In Activity 6, students apply what they have learned about ocean currents to predict and chart routes for real-life situations. In a class discussion, students share their ideas and routes. Teachers can use student maps and discussion participation to assess understanding. (Outcomes 1 and 2)

Current Stories/Message in a Bottle. In Activity 7, students make up stories about ocean currents. They brainstorm a class list of the most important points and concepts they think should be included. They are asked to use their stories to express the main things they have learned about the ocean, including information on ocean currents, their causes and effects; and wind, density, temperature, and salinity. Teachers can use this assignment as an assessment for the entire unit, as well as gaining insight into language arts and other creative abilities. (Outcomes 1–5)

Additional Assessment Ideas

The Planet and Us. In an intriguing "Going Further" activity for Activity 1, student groups make analogies between the planet Earth and human beings. Teachers can observe and note small group discussions and participation by students as they raise points for inclusion on the poster. (Outcome 1)

Coming Up with Key Concepts. At the end of Activity 5, students work in pairs to write their own Key Concepts that represent things they learned and want to remember. This idea can be expanded to the entire unit, encouraging students to take charge of their own learning by deciding for themselves what they think is important. The Key Concepts they come up with can become part of student portfolios used for assessment. (Outcomes 1–5)

Concept Mapping. Do a follow-up assessment with concept mapping. Provide a list of main ideas (e.g., the importance of currents, what is a current, causes of currents, role of density) and have the students make a concept map to connect these ideas together. (Outcomes 3 and 4)

Famous Voyages and Seafaring Feats. Assign students a research project to find out more about a famous voyage or other ocean and currents-related episode. (Outcome 2)

The Problem of Ocean Pollution. Assign students, as individuals or in groups, to do research on ocean pollution, find out its major sources, consider how it is spread by currents and other factors, and propose solutions. (Outcome 5)

The Impact of Global Warming. Global warming could have a profound impact on the ocean and the entire planet. Consider having students write an essay on this topic. (Outcome 2)

Literature Connections

Across the Big Blue Sea: An Ocean Wildlife Book
by Jakki Wood
National Geographic Society
Simon & Schuster, New York. 1998
Grades: K–3

In this picture book, a young boy launches a tiny red boat into the ocean off the California coast. It travels around the world to England. Along the way, the illustrations identify the ocean wildlife and natural objects the boat passes. Includes a map of the boat's journey. Although the book is below the grade level of the guide, students can use information about currents to decide whether or not the journey of the toy boat is possible.

Adrift: Seventy Six Days Lost at Sea
by Steven Callahan
Random House, New York. 1996
Grades: 7–Adult

After his small sloop capsizes only six days into his solo voyage around the world, Steven Callahan, floating in an inflatable raft, has to fight for his life. He is racked by hunger, buffeted by storms, and broiled by the tropical sun. He fights off sharks with a makeshift spear and watches nine ships pass by. This is the true story of the only man in history to have survived more than a month alone at sea.

Bounty Trilogy
by Charles Nordoff and James Norman Hall
Little, Brown and Co., Boston. 1995
Grades: 7–Adult

These three novels are classics of historical fiction and deal with events stemming from the mutiny of the HMS *Bounty.* The first, *Mutiny on the Bounty,* depicts the horrific struggle to round Cape Horn against wind and tide and the *Bounty's* final failure to do so, conditions which led to the infamous mutiny. The second volume, *Men Against the Sea,* is the incredible account of the 3600-mile voyage taken by Captain Bligh and 18 loyal men who were set adrift in the Pacific by the mutineers in the ship's long boat. The third volume, *Pitcairn's Island,* is the narrative of those mutineers who escaped capture and found refuge on an idyllic Pacific island.

By the Great Horn Spoon!
by Sid Fleischman; illustrated by Eric von Schmidt
Little, Brown and Co., Boston. 1988
Grades: 4–8

In this adventure novel about sailing from Boston around Cape Horn to the California gold rush, several passages exemplify the way density figures into the gold panning process.

Call It Courage
by Armstrong Sperry
Aladdin, New York. 1971
Grades: 3–6

Based on a Polynesian legend, this tells the story of a young boy who overcomes his fear of the sea and proves his courage to himself and his tribe. The story illustrates his culture's connection to the ocean.

The Cay
by Theodore Taylor
William Morrow, New York. 1991
Grades: 6–8

When the freighter on which they are traveling is torpedoed by a German submarine during World War II, Phillip, an eleven-year-old white boy, blinded by a blow on the head during the explosion, and an old West Indian named Timothy are cast up on a very small Caribbean island (known as a cay). Phillip must adjust to his blindness, overcome his prejudice, and understand the dignified, wise, and loving old man. *Timothy of the Cay*, the sequel to this book, is also recommended.

Darwin and the Voyage of the *Beagle*
by Felicia Law; illustrated by Judy Brook
Andre Deutsch, Great Britain. 1985
(distributed by E.P. Dutton, New York)
Grades: 4–8

The story of a cabin boy who goes along on Charles Darwin's famous five-year voyage. He assists Darwin with his collections of insect, bird, and marine life specimens. The format is oversized, with many drawings, charts, and maps.

***Endurance*: Shackleton's Incredible Voyage**
by Alfred Lansing
Carroll & Graf, New York. 1999
Grades: 7–Adult

A fast-paced and thrilling chronicle of Shackleton's epic Antarctic survival adventure.

Into the A, B, Sea: An Ocean Alphabet
by Deborah Lee Rose; illustrated by Steve Jenkins
Scholastic, New York. 2000
Grades: K–3

The delightful rhyming text combine with the vivid cut-paper illustrations to give the reader a tour of the ocean and its inhabitants—from Anemone to Zooplankton. Each verse succinctly captures its creature's unique attribute— "…where kelp forests sway and leopard sharks prey…" A glossary provides further information on each animal, and a teacher's supplement is available. Although intended for a young audience, this book is useful for showing about the variety of marine organisms.

Island of the Blue Dolphins
by Scott O'Dell; illustrated by Ted Lewin
Houghton Mifflin, Boston. 1990
Grades: 5–12

Left alone on a beautiful but isolated island off the coast of California, a young Native American girl spends 18 years, not only merely surviving through her enormous courage and self-reliance, but also finding happiness in her solitary life. Interwoven are descriptions of the island, fish and ocean vegetation, animals, and plants. A particularly appealing aspect of the story is the way she interacts with nature to survive, hunt, build shelter, and design clothing, both as she had been taught by her people and as she develops her own skills.

The Magic School Bus On the Ocean Floor
by Joanna Cole; illustrated by Bruce Degen
Scholastic, New York. 1992
Grades: 4–6

In her own predictable style, Ms. Frizzle takes her class on a field trip to the ocean (though the students expected a trip to the beach). The class explores many different ocean habitats and learns about the organisms in each. In one of the reports along the edge of the page, a student discusses the "rivers" in the ocean—the ocean currents.

Get Connected–Free!

Get the *GEMS Network News*,

our free educational newsletter filled with…

- **updates** on GEMS activities and publications
- **suggestions** from GEMS enthusiasts around the country
- **strategies** to help you and your students succeed
- **information** about workshops and leadership training
- **announcements** of new publications and resources

Be part of a growing national network of people who are committed to activity-based math and science education. Stay connected with the* GEMS Network News. *If you don't already receive the* Network News, *simply return the attached postage-paid card.

For more information about GEMS call (510) 642-7771, or write to us at GEMS, Lawrence Hall of Science, University of California, Berkeley, CA 94720-5200, or gems@uclink4.berkeley.edu.

Please visit our web site at www.lhsgems.org.

GEMS activities are effective and easy to use. They engage students in cooperative, hands-on, minds-on math and science explorations, while introducing key principles and concepts.

More than 70 GEMS Teacher's Guides and Handbooks have been developed at the Lawrence Hall of Science — the public science center at the University of California at Berkeley — and tested in thousands of classrooms nationwide. There are many more to come — along with local GEMS Workshops and GEMS Centers and Network Sites springing up across the nation to provide support, training, and resources for you and your colleagues!

Yes!

Sign me up for a free subscription to the

GEMS Network News

filled with ideas, information, and strategies that lead to Great Explorations in Math and Science!

Name____

Address______________________________________

City__________________ State______ Zip________

How did you find out about GEMS? (Check all that apply.)

❒ word of mouth ❒ conference ❒ ad ❒ workshop ❒ other: ______________

❒ In addition to the *GEMS Network News*, please send me a free catalog of GEMS materials.

GEMS
Lawrence Hall of Science
University of California
Berkeley, CA 94720-5200
(510) 642-7771

Ideas
Suggestions
Resources

Sign up now for a free subscription to the *GEMS Network News!*

that lead to Great Explorations in Math and Science!

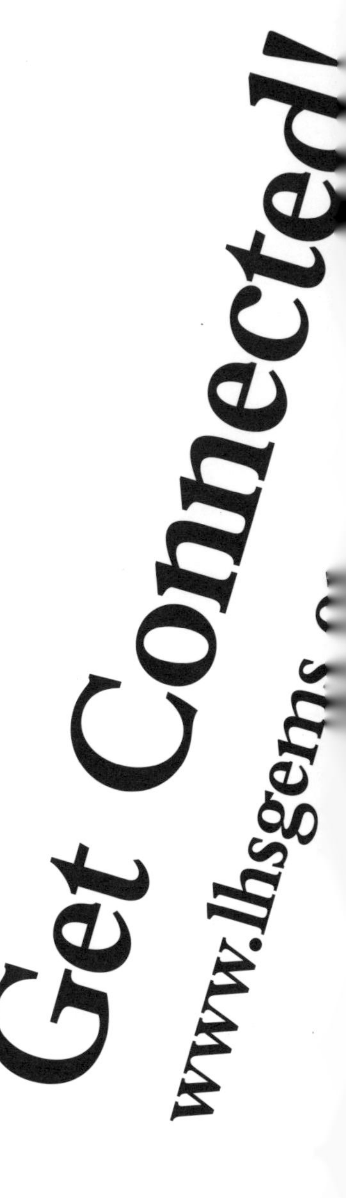

101 LAWRENCE HALL OF SCIENCE # 5200

1-61571-25775-62-X

NO POSTAGE NECESSARY IF MAILED IN THE UNITED STATES

BUSINESS REPLY MAIL
FIRST-CLASS MAIL PERMIT NO 7 BERKELEY CA

POSTAGE WILL BE PAID BY ADDRESSEE

UNIVERSITY OF CALIFORNIA BERKELEY
GEMS
LAWRENCE HALL OF SCIENCE
PO BOX 16000
BERKELEY CA 94701-9700

Moby Dick
by Herman Melville
Viking Penguin, New York. 1996
Grades: 7–Adult

The full version of this classic novel of the obsessed Captain Ahab may prove daunting for your students. The abridged version removes the sections that detail life aboard a whaling ship and concentrates on the fate of the *Pequod*. That said, Ahab's tracking of Moby Dick follows ocean currents nearly around the world.

Out of the Ocean
by Debra Frasier
Harcourt Brace and Company, San Diego. 1998
Grades: K–5

As a young girl and her mother walk along an Eastern Florida beach, they marvel at the many treasures cast up by the sea and the wonders of the world around them. Detailed and illustrated pages at the end of the book give information about the items found. One of the items is a note-filled bottle and ocean currents are discussed. Although written for younger students, the book is excellent for showing, through photographs and collages, the diversity of ocean life as found strewn on the beach by currents, waves, and the tides.

The Robinson Crusoe
by Daniel Defoe
Penguin Putnam Books for Young Readers, New York. 1995
Grades: 7–Adult

Consistently popular since its first publication in 1719, this is the classic story of a young man who sets sail for a life of adventure in far away places. Fleeing pirates, he is swept ashore on a deserted tropical island and must learn how to survive and deal with his isolation.

Treasure Island
by Robert Louis Stevenson
Penguin Putnam Books for Young Readers, New York. 1988
Grades: 7–Adult

Originally published in 1883, this is the classic tale of adventure on the high seas and the search for buried treasure featuring such characters as the young and honest cabin boy Jim Hawkins, the sinister Israel Hands, and the hero-villain pirate Long John Silver.

The True Confessions of Charlotte Doyle
by Avi
Avon, New York. 1990
Grades: 5–8

In the summer of 1832 aboard a ship crossing the Atlantic from England to America, 13-year-old Charlotte Doyle—the only passenger on the ship—finds herself caught between a ruthless captain and a mutinous crew. This book is her account of that voyage which tests her courage and her will to survive.

The Voyager's Stone: The Adventures of a Message-Carrying Bottle Adrift on the Ocean Sea
by Robert Kraske; illustrated by Brian Floca
Orchard Books, New York. 1995
Grades: 3–6

A message-carrying bottle, thrown into the sea by a boy vacationing in the Caribbean, makes its way eastward in the Atlantic, then south to Antarctica, and onward to Australia, where it is discovered by an Aborigine girl. Through the bottle's voyage, oceanography is explored—covering such topics as currents, animals, and the variety of life found along the margins of the world's oceans.

Windcatcher
by Avi
William Morrow, New York. 1992
Grades: 4–7

While learning to sail during a summer visit to his grandmother's house on the Connecticut shore, 11-year-old Tony becomes excited about the rumors of sunken treasure in the area and starts following a couple who seem to be making a mysterious search for something.

The Wreck of the Whaleship *Essex*, A Narrative Account
by Owen Chase, First Mate;
edited by Iola Haverstick and Betty Shepard
Harcourt, Brace & World, New York. 1965
Grades: 7–Adult

A firsthand account of the tragedy of the whaler *Essex* if somewhat self-serving.

Summary Outlines

Activity 1: Planet Ocean

Getting Ready

1. Plan strategy for gathering photos and ocean images.
Plan purchase of posters and inflatable globes.
2. Tape up chart paper for Ocean Brainstorm. On another strip of paper write "OCEAN." Write out the Key Concept:
- **There is only ONE ocean! Our Earth is covered by one interconnected world ocean that circulates around all the continents.**
3. Hang posters. Ask students to think about ocean-related ideas and experiences. Inflate globes. Display large globe.
4. Decide which worksheet to use for Session 2 and duplicate.

Session 1: Ocean Brainstorm

1. Arrange students in groups. Show videotape, slides, or play audio tape.
2. Point out posters and chart title. Each group will brainstorm about the ocean. Groups discuss for a few minutes.
3. Hold up ocean pictures. Distribute pictures and have student groups talk.
4. Regain attention of class. One person in each group writes down at least five things about the ocean. Each person contributes at least one thing.

Organizing the Information

1. Tell students they will organize what they know about the ocean.
2. Have volunteers share something they recorded. Create diagram with categories. Place "The Physical Ocean" at the center.
3. If fact could go in more than one category, students help decide placement. Use arrows to show connections. Record questions.
4. Acknowledge that students know a lot. Say this unit focuses on the physical ocean. Use arrows to connect items to physical ocean. See if they can add more about water and currents.
5. Ask students to relate items in the physical ocean to other categories until students realize everything relates to the physical ocean.

Session 2: Global Exploration

1. Give one globe to each group and let them freely explore.
2. Using globes, review latitude and longitude. Hand out Global Exploration student sheets, paper, and markers. Introduce the term ***ocean basin***.

Discussing the Global Exploration Worksheet

1. After students complete Questions 1–5, help class analyze. **Emphasize:** Water covers more of Earth's surface than land (about three quarters); Southern Hemisphere is about 80% water, Northern Hemisphere about 60%; on many maps it appears there's more land, but Earth should be called "Planet Ocean." The ocean is the most important feature on the surface of the planet, distinguishing ours from other planets in solar system.
2. Discuss Question 6. **Emphasize:** Much of world's food supply comes from the ocean; coastlines are places where ideas and goods are exchanged; much of history relates to winds and currents.
3. Discuss Questions 7–9. **Emphasize:** The view of the Southern Ocean shows Southern Hemisphere doesn't have much land; Antarctica is relatively small; there is only one ocean.
4. Discuss Question 10. **Emphasize:** Red arrows are warm water currents, blue arrows cold; ocean currents are huge amounts of water moving in a particular direction; ocean currents move water throughout the world ocean.

Concluding the Discussion

1. Tell class they will focus on ocean currents for the next few sessions.
2. Hold up Key Concept, have it read aloud, discuss, then post.

Activity 2: Waste Disposal

Getting Ready

For the Student Explorations

1. Set up materials with some set aside for use later.
2. Cover the tables with newspaper.
3. Set up a location for students away from the materials.
4. Obtain ice cubes and find a way to keep them cold.
5. For each group, mark an "X" on the Waste Disposal data sheet to represent their "country." Assign a different country to each team.

For Teacher Demonstration

1. Have five straws on hand for students to create wind. Make the colored confetti. Make transparency of Pacific Rim Map.
2. Tape chart paper to wall. Project map overhead. Draw map outline. Draw in equator, but don't add currents. Remove the map transparency.
3. Place salad container on overhead. Its left and right edges should line up with Asia and the Americas.
4. Fill the two containers three-quarters full of water.
5. Practice making wind, gyres, and wind with an obstacle before class.

6. Write out Key Concepts:
- **Things dumped into the ocean may be distributed by currents throughout the ocean.**
- **Wind and the temperature differences between masses of water are two factors that cause currents.**
- **Winds blowing across the surface of the ocean—combined with other factors—cause major circulating currents, or gyres.**

Session 1: Student Explorations

Thought Swap

1. Tell students they will get a chance to talk with different classmates. Each partner should be a good listener and speak clearly in turn.
2. Have students stand shoulder to shoulder in parallel lines, with side-by-side students at least six inches apart. You will be asking a question or raising an idea for them to talk about with the person across from them.
3. Pose first question. When you call time, have a few students report something their partner told them.
4. Have one line move a position to left; person at end walks to beginning. Ask: When and where was the last time you went swimming or wading? Did you feel a current? What is a current and how could you tell a current was happening? Where and when have you noticed cold and warm layers of water? Describe ways to make a current in a pool, tub, or glass of water.
5. Write the last question on chart paper and record answers for later reference. Create groups of four for next activity.

Introducing the Activity

1. Ask students if they know where garbage and wastes are dumped. If no one mentions the ocean, do so.
2. Tell students they are going to simulate and track the dumping of wastes at sea.

Explaining the Procedure

1. Explain procedure while demonstrating on overhead.
2. Tell students their "country" has decided to dispose of waste in the ocean. Their job is to find the best and worst dumping locations.
3. Students decide on four test sites, using different food coloring. They use four drops—except eight at yellow site. Each color is monitored by a team member who records direction of current.
4. Demonstrate dropping colors at sites you've chosen. Ask for observations. Ask what is missing [ice cubes]. Explain that you are modeling doing a control and discuss.

5. Tell them there will be no wind at first , but later there will be. Wind direction is marked on maps and they may consider it when choosing dumping sites.
6. Remind students not to jiggle or blow on tray. They place ice cube at location shown and drop coloring at the four sites they select.

Placing and Tracking Waste in the Ocean

1. Ask students if they have questions. Have students set up their materials. Distribute coloring and ice cubes. As they work, sketch Waste Disposal sheet on board.
2. Have teams combine results from all four sites onto chart paper, with "X" marking worst location and "O" the best.
3. Distribute a Student Explorations with Waste Disposal sheet to each team. Have them choose a recorder.

Discussing the Results

1. After 15–20 minutes, teams record on board their best and worst locations.
2. Gather students away from materials. Ask for observations and questions, and how they might find out more.
3. Ask how well the model simulates the ocean. What are the limitations? How could it be better?
4. What generalizations might they make about how currents are formed? How could generalizations be tested with this model and in the ocean?
5. Encourage students to draw their ideas on the board. Discuss best and worst locations and if "continents" affected the currents.
6. If students say best location is where it is least likely for the waste to reach *their* country, ask if that's fair and encourage discussion.

Introducing Wind

1. Tell students another factor will be added—wind. One person at a time will blow "wind" through a straw only in the direction shown on the map. They should observe but not draw their results on the map.
2. Pass out four straws to each team.
3. When most food coloring has been spread in the trays collect the materials and ask what happened.
4. Say this is similar to something injected into a person's bloodstream. It can take a long time, but things dumped may be distributed by currents throughout the ocean.
5. Remind students there is only one ocean and that water of different temperatures and wind are two current-causing factors. Hold up, have read out loud, discuss, and post the first Key Concept.

Session 2:
Demonstrating Wind-Driven Currents

1. Tell students you are going to simulate wind blowing over the ocean at the equator. You will need five volunteers. Direct overhead over Pacific Rim Map. Have students copy the outline and equator.
2. Have first student blowing through a straw create an east-to-west current on the projected image. The straw should be **level with the surface** and held far enough back to blow **lightly** along equator.
3. Have students record what they observe, with labels and arrows for direction of water and wind. Lead discussion about what they observed.
4. As needed, you can have other students take over blowing.
5. This model simulates how winds set the surface water flowing in the same direction as wind. Winds near equator (the trade winds) generally blow from east to west, causing east to West Equatorial Current. Draw this current on the Pacific Rim Map and have students label it on their drawings.

Demonstrating Gyres

1. Tell students that if there were no land, but only ocean, the Equatorial Current might flow east to west all the way around the globe.
2. The next part of the demonstration models what happens when the Equatorial Current runs into continents.
3. Drop 10–15 plastic pieces into pan and ask students to observe.
4. Have a student hold the end of the straw down low and level with the side of the pan, so wind from the straw goes across the equator as before. You may have to make adjustments to get current centered so the plastic pieces deflect both right and left to form gyres.
5. Have class draw what they observe on their own map. Ask students to describe what they observed and have a student draw a diagram.
6. If not mentioned, point out two circular currents, one clockwise in the Northern Hemisphere, the other counterclockwise in Southern Hemisphere. These are gyres. Students should add them to their drawings.
7. Ask students whether currents reaching west coasts of North America and South America would be warm or cold.
8. Show map overhead of currents in Pacific or pass out copies to students to compare currents in the model with the ocean.

Introducing an Obstacle

1. Move overhead so it projects onto blank paper. Ask students to reflect on waste disposal activity. How were wind-driven currents affected by continents? How will a current be affected by an island in its path?

2. Draw a rock outline on the chart paper or board. Have students discuss and/or draw their predictions.
3. Have a student direct wind in direction of island (rock). Eddies and countercurrents form. Point out how complicated currents can become where there are many obstacles.
4. Challenge students to find gyres on their globes. There are many islands and irregular shapes. Other factors also complicate current patterns.
5. Ask a team to share one gyre. Which other teams found the same gyre.? Continue with other gyres, pointing out where obstacles have created strange currents, and where gyres have been altered.
6. Discuss the limitations and advantages of models.
7. Hold up, have read aloud, discuss, and post the Key Concepts.

Activity 3: Current Trends—Station Rotations

Getting Ready

1. Collect plastic bottles for Stations 1 and 2.
2. Determine how to make and keep hot and icy cold water.
3. Arrange and have ready materials for each of the stations. Make a label for each station.
4. Make two copies of Student Station Directions for each station on colored paper. Duplicate Prediction data sheet and Current Trends data sheets for each student, and World Map of Currents for each group.
5. Write out debriefing questions on chart paper.
6. Write out the Key Concepts:

- **Salinity and temperature differences create masses of water with different densities.**
- **Gravity causes more dense water to sink below less dense water. As a result, the less dense water rises.**

Session 1: Introducing the Stations and Making Predictions

1. Tell students they will be rotating in small groups through current-related stations. There are two identical sets of stations, A and B. Briefly demonstrate or describe setup of Station 1.
2. Distribute Prediction data sheet and have materials available. Discuss predictions; show how to draw one on data sheet. Have students make predictions for Station 1 and illustrate.
3. Demonstrate or describe setup for Stations 2 and 3, pausing after each for students to illustrate predictions.

Station Rotations

1. Assign students to groups, and groups to stations.
2. Pass out Current Trends data sheet, one per student. One of them should read the station directions aloud. Ask students to divide up into roles needed.
3. When you say to rotate they go to the next station—those at Station 3 go to Station 1. Give students 15 minutes to complete each of two stations.

Session 2: Completing the Stations

1. Have students complete their third (final) station.
2. Tell the groups they'll prepare a poster about this station to present to the class.

Making Station Posters

1. Tell students they will discuss results of the stations to relate them to the real world. They will prepare a poster.
2. Distribute World Map of Currents and globes to groups. Also distribute chart paper and markers.
3. Tape Station Summary chart on board and have the recorder in each group write their responses on the poster.
4. Have groups tape up posters around the room, give a short summary, and respond to questions.
5. Leave posters up and tell the students they might want to refer to them during class discussion.

Debriefing the Station Posters
Salinity Currents and Layering

1. Read observations and Key Concept from posters.
2. Ask students what they think causes the water to move in salinity currents.
3. Where might salinity currents occur? Why?
4. Ask students to find the Mediterranean Sea. Where might this salty water go? Although the water is salty, it is also warm. Where do they think it would end up in the Atlantic?
5. Have students find the Amazon River, and where it flows into the ocean. Ask what might happen to all that fresh water when it reaches the ocean.

Temperature Currents and Layering

1. Read the observations and Key Concept from the student posters. Ask students what they think causes the water to move in temperature currents.
2. Have they ever noticed when swimming how surface waters are warmed by sun and the depths cold? When is this most evident?
3. Ask why warm water remains on top, and cold water sinks below.

4. What time of year might cold water be found on top, and warmer water below? Will the water stay layered this way? Where might temperature currents occur? Why?
5. When salinity and temperature differences combine, different water masses don't mix easily, but flow above or beneath each other.
6. How might these layers affect ocean animals?
7. Explain that upper layers are where plantlike organisms live, because they need sunlight, but that nutrients from decomposed organisms sink to the bottom and collect in the colder and denser water. If these nutrient-rich waters are brought back to the surface, they can create areas rich with life.

Polar Versus Tropical Water
1. Read the observations and Key Concept from the student posters. Ask where they think the phenomenon shown by the station might occur. Why?
2. Although layers in the ocean may sometimes last a long time, there is also a lot of mixing. What may cause layers to mix?
3. Did anyone notice dense, cold water move along the bottom then hit the side of the container and flow back up? This is called ***up-welling.*** Why is this important to organisms? Upwelling zones are usually very rich with life.
4. Winds are usually the cause of coastal upwelling by moving the surface waters offshore. Most wind-driven currents occur in top 20% of the ocean. The other 80% is moved mainly by density differences.
5. Hold up, have read out loud, discuss, and post Key Concepts.

Activity 4: Layering Liquids

Getting Ready

Preparing the Solutions
1. Write and post a Key to the Colors.
   red—hot and salty
   blue—cold and salty
   yellow—hot and fresh
   green—cold and fresh
2. Label the four thermoses and add about 30 drops of the appropriate food coloring to each thermos.
3. Add kosher salt to containers:
   red = 14 level tablespoons (about ¾ cup) salt
   blue = 14 level tablespoons salt
   yellow = no salt
   green = no salt

4. Have ice cubes available. Measure and add two cups of room temperature or cold water and four ice cubes each to the thermoses labeled:
  blue—cold and salty
  green—cold and fresh
5. Close cold-water thermoses to prevent warming. Shake one vigorously for about 30 seconds to dissolve salt.
6. Heat about six cups of water. Measure and add two cups of hot water each to the thermoses labeled:
  red—hot and salty
  yellow—hot and fresh
7. Close hot-water thermoses to prevent cooling. Shake one to dissolve salt. Shake thermos again for about 30 seconds to finish dissolving salt.

For each table

1. Cut potatoes in slices. Have extras. Cut two straws in half for each student pair. Have extras.
2. Duplicate Density Layers Plan for each student and Density Layers data sheet for each pair.
3. Label insulated cups: blue, red, yellow, green.
4. Prepare one equipment tray per group, with one dropper in insulated cup.
5. Prepare demonstration materials on tray. (Fill two cups with same amount of water, about three-quarters full. Fill two other clear plastic cups three-quarters full of marbles.)
6. Make about a tablespoon of purple water.
7. Arrange tables so two or three pairs of students can work together. Surfaces should be as level as possible.
8. Place trays, extra straws, purple water, oil, and thermoses in central location.
9. Write out the Key Concepts:

- **The ocean is made up of layers of waters of different densities.**
- **Cold water is denser than warm water.**
- **Water with salt is denser than fresh water.**
- **The more closely packed the molecules in a substance, the denser the substance.**

Session 1: Introducing the Challenge

1. Tell students they will try to create layers of colored liquid using only water and salt. Ask about salt water compared to fresh water—if necessary, remind them salt water is more dense. What about cold versus hot? As needed, remind them cold water is more dense.
2. Show materials; challenge students to make as many layers as they can. Also show them the liquids and the Keys to the Colors chart. Tell students the color key is also on Density Layers data sheet. Show how to color in and label predictions and results, circling sequences that layer successfully.

3. Distribute Density Layers Plan. Each student makes a plan then compares it with partner's plan. Together they decide what to do and why. If first plan doesn't work, they can try others, but must write it down first.

Demonstrating the Procedure

1. Caution students not to taste any of the liquids.
2. Insert the straw into a slice of potato at a 45° angle. Add purple water down the straw, then add oil. Hold piece of white paper behind straw. Show how to empty the contents of the straw.
3. Show the students where the thermoses are. If using hot water, students should get a small amount in their insulated cup before using it, so it will still be hot.
4. Stress the challenge: to layer two or more liquids in the straw so none mix together. Remind them to record predictions.
5. Demonstrate "sloppy" methods to show results if layers *do* mix.
6. Show how, after each test, partners should draw and label sequences they tried on data sheet, circling successful attempts.

Testing

1. Divide class into groups and distribute material. At each table have students divide into teams of two.
2. Distribute Density Layers Plan. After they complete it, have them work together in pairs on a team plan.
3. Give Density Layers data sheet to each pair and have them begin. When most teams have layered at least two liquids, have them return materials.

Discussing Results

1. Tell students to put up data sheets. Ask for observations. Is there any color that shows up at bottom most often? Why? Is there a color that frequently was top layer? Why?
2. If any made four layers, draw attention of class to data sheet. Ask them to rank from least to most dense. If no one was able to make four layers, ask class to help rank them in the order that they think they would layer in.
3. Why did liquids form layers? How might students make more layers? Ask students about situations where liquids layer in the ocean.

Session 2: Discussing Density

Salt Water and Fresh Water: Demonstrating Density

1. Show students two cups of water—you measured same amount into each. Show them two cups of marbles, and tell students to imagine they represent two cups of water at molecular level.

2. Density relates to how tightly packed molecules are together—the tighter the pack, the more dense.
3. Have students imagine that each of the two marble cups has the exact same amount of water measured into it, like the cups of water. Which would be densest?
4. Add two tablespoons of salt to one of the cups of water. Stir and say this is how salt solutions were made. Review which is more dense: the cup with only water or the cup with water and salt?
5. They'll now see how this looks in the molecule model. Pour some salt into one of the cups with marbles. Ask what they observe.
6. Salt and water molecules mix in a similar way, making the mixture denser. Does the cup with marbles and salt weigh more than the other cup?

Introducing Temperature and Density

1. Say it's easy to see how, if salt is added, water becomes denser—but what about water without other ingredients? Why is cold more dense than hot?
2. Molecules are always moving. They bump into each other and push each other away. The hotter a liquid, the faster molecules move, the more they hit each other, the more space between them, and that makes the liquid less dense. Cold water still has same amount of molecules, but they move less, are able to pack together more, so the liquid is more dense.

Density and Currents

1. Brainstorm where these four types of water might be found in the ocean.
2. After these waters enter the ocean, they form layers which may last a long time and move long distances.
3. Students made stable layers, but in ocean, waters of different densities move, causing currents. Did they see a "mini-current" in straws? How would waters of different densities affect currents? Discuss how warm water rising and cold water sinking can cause currents, as can saltier water sinking and less salty water rising.
4. Ask students to read out loud, discuss, and post the Key Concepts.

Activity 5: Ice Cubes Demonstration

Getting Ready

1. Fill two identical jars about three-quarters full with tap water. Add about a quarter cup of kosher salt to one jar, mix thoroughly, and let sit.

2. Display Key Concepts for Activities 1–4; student responses to questions posed in Thought Swap in Activity 2 and student posters from Activity 3.
3. Write questions on What Happened? chart. With hot and cold water, which temperature water ended up at bottom? With salty and fresh water, which one ended up at bottom? In which jar did ice cubes melt fastest?
4. Make a copy of the Think, Pair, Share student sheet for each student.

Making Predictions

1. Show students two jars filled with water. One is salt water and the other fresh (don't tell which is which).
2. How could they figure out which is which? Have them jot down at least two ideas and keep notes to add to later.
3. Discuss their ideas, then say that while many of their suggestions would probably work, we can also tell by using ice cubes.
4. Ask students to predict if ice cubes will melt faster in fresh water or salt water, write their prediction, and explain their reasoning. Discuss, then take a show of hands to gauge class opinion.

Ice Cubes Demonstration

1. Have a couple of students add 3–4 ice cubes to each jar (the same amount to each). Tell students not to bump or disturb the jars.
2. Have students draw and label the setup and record observations with words and diagrams on paper used in Making Predictions.
3. Have a couple of students look in the jars and report where ice is melting faster.
4. After ice cubes melt some, add 3–4 drops of food coloring to each jar. Place a sheet of white paper behind the jars. Ask students to record observations.
5. This may help explain which ice cube melted faster.
6. Tape up the What Happened? chart. Ask students to write results.
7. Lead a discussion, writing responses on chart. People often get confused because of other experiences. Results have to do with densities, not with freezing temperatures.
8. Ask students to use observations to explain results.

Discussing Explanations

1. Have students present statements and have the class analyze them. Discuss. If students explain, briefly summarize. If students are unclear, ask Guiding Questions to clarify.
2. Explain that the densest water in the ocean is formed around Antarctica, because the water is very cold and very salty. It sinks to the bottom of the ocean basin surrounding Antarctica and travels north.

Think, Pair, Share, or Revisiting Thought Swap

1. Tell students they are going to do a Think, Pair, Share activity to check their understanding of the concepts they've been investigating.
2. Distribute Think, Share, Pair student sheets for responses to: What is an ocean current? What sets water in motion and causes ocean currents to form? Ways to make a current in a pool, tub, glass of water? Why do you think ocean currents might be important?
3. Have them share their responses with a partner, then with a small group. Students can add to their lists.
4. Lead discussion and record ideas. Refer to Key Concepts as appropriate.
5. Refer back to chart paper from Activity 2 to compare what they now know about "ways to make a current in a pool, tub, or even a glass of water."

What is the Key Concept?

1. Have small groups of students write/present their own Key Concept.
2. Lead a discussion comparing concepts, come to a consensus, and display.

Activity 6: Ocean Routes

Getting Ready

1. Make copies of student maps and make overheads. Decide if you will save some scenarios as homework.
2. Make copies of station sheets. Set them up as stations, with designated colored pens.

Introducing the Stations

1. Ask students to brainstorm real-life situations where knowledge of currents is important.
2. Show Surface Currents overhead—students will use this map. Remind them of gyres. Show Deep Currents overhead and remind them of upwelling zones, downwelling zones, and deep currents. Students will also use this map.
3. Tell students routes they choose do not have to match actual routes, but they do need to follow currents.
4. Assign students to first station. Have them move through stations at their own pace. Distribute student sheets and pencils and allow students to begin.
5. End when most students have completed the stations.

Debriefing the Stations

1. Go through questions on station sheet.
2. For each station, ask a few students to share routes chosen. Compare with currents on map. Show actual route and share information in Station Debriefings.

Activity 7: Message in a Bottle

Getting Ready

1. Have current maps and writing materials on hand.
2. Post charts, posters, and Key Concepts from the unit.
3. Decide if there is anything you want students to include in their stories and whether to assign as homework or write in class.

Introducing the Activity

1. Remind students of ocean current stories. Have they heard others? Today they will make up their own.
2. Point out charts, posters, and concepts on display. Have student groups take a "gallery walk," with a recorder noting the 10 most important ideas.
3. After groups make their list, brainstorm and record ideas. Have class help highlight the most important points to incorporate.

Clarifying the Task

1. Stories can be about many current-related subjects, in various literary forms, but must have accurate information about ocean currents.
2. Show current maps on overhead. Students need to draw the route from their story on the map, and attach it to their story. Give an example or two.

Creating the Stories

1. Distribute maps and have students begin.
2. Have students share their stories.